# Durability of Reinforced Concrete from Composition to Protection

Carmen Andrade • Joost Gulikers • Rob Polder
Editors

# Durability of Reinforced Concrete from Composition to Protection

## Selected Papers of the 6th International RILEM PhD Workshop held in Delft, The Netherlands, July 4-5, 2013

Springer

*Editors*
Carmen Andrade
Institute of Construction Science
Eduardo Torroja (IETcc)
Madrid, Spain

Joost Gulikers
Ministry of Infrastructure
and The Environment – GPO
Utrecht, The Netherlands

Rob Polder
Delft University of Technology
Delft, The Netherlands

ISBN 978-3-319-09920-0 ISBN 978-3-319-09921-7 (eBook)
DOI 10.1007/978-3-319-09921-7
Springer Cham Heidelberg New York Dordrecht London

Library of Congress Control Number: 2014954127



Printed on acid-free paper

Springer is part of Springer Science+Business Media (www.springer.com)

# Contents

# Visualisation of the Electrical Resistivity Distribution of Reinforced Concrete

**Kenji Reichling, Michael Raupach, and Norbert Klitzsch**

**Abstract** The electrical resistivity of concrete is a characteristic property correlating with different durability related parameters, e.g. the degree of water saturation, the corrosion rate of the reinforcement or the chloride diffusion coefficient. On structures the resistivity can be determined non-destructively by using a Wenner setup. The analysis is based on the assumption of a homogeneous semi-infinite concrete element. Based on the heterogeneous structure of reinforced concrete (e.g. reinforcement bars, moisture differences) the electrical field is deformed which may lead to misinterpretations.

To take these influences into account the application of the Electrical Resistivity Tomography (ERT) on reinforced concrete is studied. The method is based on several measurements with different electrode arrangements and distances in order to determine the true resistivity distribution in the concrete. The aim of the application is to identify local heterogeneities in order to quantify the resistivity distribution in concrete accurately.

**Keywords** Reinforced concrete • Non-destructive testing • Electrical resistivity • Electrical resistivity tomography • Wenner • Dipole-dipole • Sensitivity

## 1 Introduction

The electrical resistivity of concrete is playing an increasingly important role in the area of condition surveys on concrete structures. In the past correlations with several durability related parameters e.g. corrosion rate, water content, porosity and chloride diffusion coefficient could be shown (Weydert and Gehlen 1999; Raupach 2006; Castellote and Andrade 2006; Andrade et al. 2009; Warkus and Raupach 2010). The exact correlations are not completely clarified until now but the relevance of the electrical resistivity of concrete is undisputed.

K. Reichling (✉) • M. Raupach
Institute of Building Materials Research, RWTH Aachen University, Aachen, Germany
e-mail: reichling@ibac.rwth-aachen.de

N. Klitzsch
Institute for Applied Geophysics and Geothermal Energy, RWTH Aachen University, Aachen, Germany

 
C. Andrade et al. (eds.), *Durability of Reinforced Concrete from Composition to Protection*, DOI 10.1007/978-3-319-09921-7_1

Principally the resistance $R$ is determined by Ohm's law:

$$R = \frac{U}{I} \tag{1}$$

where $I$ is an applied current and $U$ the resulting voltage. Basically the resistance is influenced by the electrode arrangement and the geometry of the electrical field. Therefore the geometry independent resistivity $\rho$ is used as material property, whereas the geometric factor $k$ accounts for the electrode positions:

$$\rho = k \cdot R \tag{2}$$

For resistivity measurements two and four electrode setups are typically used (Polder et al. 2000). To counteract the influences of polarised electrodes a four-electrode setup is suitable in which different electrode pairs are used for applying a current and for measuring the resulting potential drop. In condition surveying the electrode configuration according to Wenner (see Fig. 1) is typically applied (Wenner 1915).

The four electrodes are aligned with an equal distance $a$. The current is injected via the electrodes $A$ and $B$ whereas the potential is measured between the electrodes $M$ and $N$. The geometrical factor is based on the consideration of a potential propagation originating from two point electrodes ($A$, $B$) in a homogeneous half space (Wenner 1915). In this case the geometric factor depends on the electrode distance $a$:

$$k = 2 \cdot \pi \cdot a \tag{3}$$

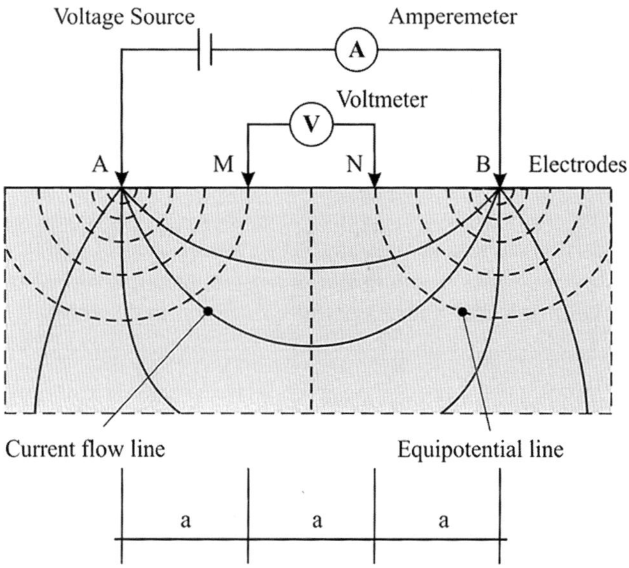


**Fig. 1** Electrode configuration according to Wenner (Wenner 1915)

Due to the pronounced heterogeneity of reinforced concrete the electrical field can be deformed, so that these assumptions and thus Eq. 3 are no longer valid. This can be the case if reinforcement bars are near the current electrodes or the resistivity distribution in the concrete is inhomogeneous, e.g. due to wetting, drying or carbonation. Furthermore the results can be influenced in the case of compact elements, e.g. columns, where the assumption of a half space is no longer valid. In these cases an apparent resistivity value is determined which is a weighted average of the true resistivity distribution.

## 2 Electrode Configurations

The real resistivity distribution inside concrete can be determined by the Electrical Resistivity Tomography. ERT allows to distinguish between concrete areas with different degrees of saturation and to identify the areas strongly influenced by the reinforcement. The method developed and frequently used in geophysical applications (e.g. Lines and Treitel 1984; Viergever and Todd-Pokroper 1988; Friedel 2000) requires measurements with many electrodes arranged e.g. in a line. Based on the measured apparent resistivity values the true resistivity distribution can be determined using iterative inversion techniques.

By varying the horizontal position of the electrode setup (see Fig. 2a) heterogeneities parallel to the concrete surface, e.g. a reinforcement bar, can be detected. Information about inhomogeneities perpendicular to the surface, e.g. humidity profiles, can be gained by varying the electrode distances (see Fig. 2b). Beside the Wenner configuration (Fig. 1), which is also called Wenner Alpha, other configurations exist, e.g. Wenner Beta and Gamma, Schlumberger or Dipole-Dipole, which are accompanied with different advantages and disadvantages. The diverse configurations are illustrated in Fig. 3.

Different electrode configurations differ in their sensitivity distributions (Barker 1979). The concrete underground can virtually be discretised into a mesh of small

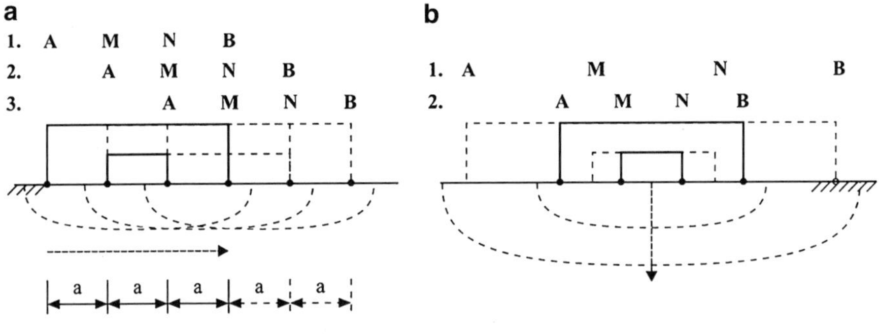


**Fig. 2** Measurement of different areas by changing the electrode positions (**a**) and measurements with different depths of investigation (**b**)

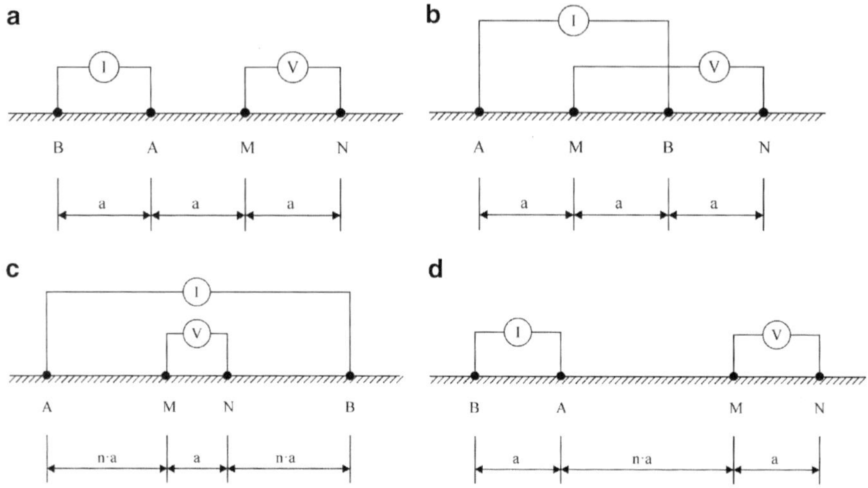


**Fig. 3** Illustration of the Wenner Beta (**a**), Wenner Gamma (**b**), Schlumberger (**c**) and dipole-dipole (**d**) electrode configurations

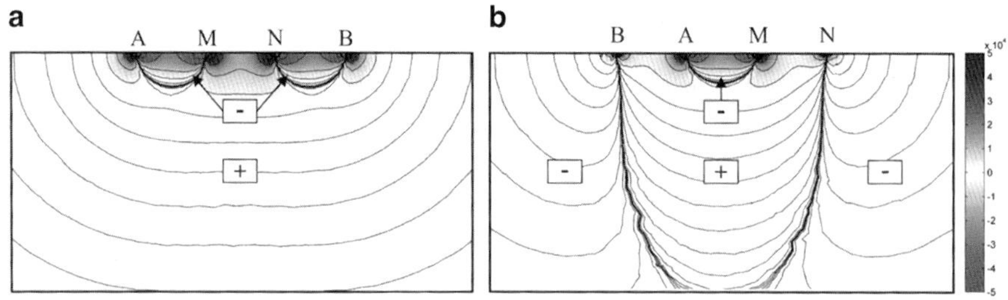


**Fig. 4** Sensitivity distribution of the Wenner Alpha (**a**) and the dipole-dipole configuration (**b**) for an area with homogeneous resistivity

elements. Even if the resistivity of the underground is homogeneous, each of the elements has an individual influence on the measured resistivity, expressed by a sensitivity value. The sensitivity distributions are primarily influenced by the electrode configuration but also by the resistivity distribution. The sensitivity distributions for the Wenner Alpha and the Dipole-Dipole configuration are exemplarily shown for the homogeneous case in Fig. 4.

The dark areas show large sensitivity values indicating a high influence on the measured values. Furthermore it can be distinguished between areas with positive and negative values. A conductive object situated in an area with negative sensitivity values would paradoxically lead to an increase of the apparent resistivity. This would be the case if a reinforcement bar with a low concrete cover depth would be situated between a current and a potential electrode.

Both distributions in Fig. 4 have different advantages regarding the reconstruction of the resistivity distribution of the underground. The Wenner Alpha configuration shows a horizontally expanded area with high sensitivities which leads to a good resolution perpendicular to the concrete surface. Comparing this with the Dipole-Dipole configuration it can be noticed that the area with high and positive sensitivities has a pronounced focus perpendicular to the surface which leads to a better lateral resolution parallel to the surface and a higher depth of investigation.

The geometrical factor according to Eq. 4 must be considered to determine the apparent resistivity when using the Dipole-Dipole configuration.

$$k = \pi \cdot n \cdot (n+1) \cdot (n+2) \cdot a \tag{4}$$

Apart from the dipole length $a$ the geometrical factor depends on the dipole distance factor $n$. Both parameters are shown in Fig. 3d.

## 3 Electrical Resistivity Tomography

As already mentioned above, information about the resistivity distribution in the concrete is gained by varying electrode positions and distances using one or different electrode configurations. In order to avoid repeated movements of the electrodes a multi-electrode array with 20 electrodes was developed for the application on concrete structures (Fig. 5).

The presented results were measured with a geophysical measurement device "4 point light 10 W" manufactured by Lippmann (Rauen 2012). The device allows an individual switching of the electrodes as current or potential electrode and thus different electrode configurations.

For each reading based on four electrodes, one apparent resistivity value is obtained. One measurement consists of typically several hundred readings. In the case of a homogeneous underground the readings would correspond to the true resistivity. Otherwise the real resistivity distribution must be numerically calculated for which a boundary value problem has to be solved by inverse computation.

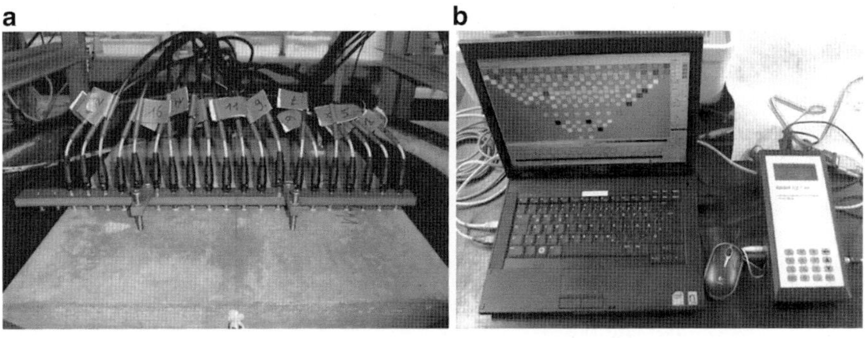


**Fig. 5** Multi-electrode array (**a**) and computer controlled measurement device (**b**)

Starting from an observed effect (potential distribution) the cause (resistivity distribution) is iteratively approximated. During the calculation process the resistivity values of the underground are adjusted stepwise and synthetic apparent resistivity values are calculated based on this data. Based on the deviation between the synthetic apparent resistivities with the measured ones new adjustment parameters are determined. The calculation is based on the Laplace equation (Eq. 5) from which the current application is considered by using the Neumann boundary condition (Eq. 6) and the potential differences by using the Dirichlet boundary condition (Eq. 7) (e. g. Lines and Treitel 1984; Viergever and Todd-Pokroper 1988; Friedel 2000).

$$-\nabla \cdot \sigma \nabla \Phi = 0 \quad in \quad \Omega \tag{5}$$

$$-\sigma \frac{\partial \Phi_{AB}}{\partial n} = j_{AB} \text{ on } \partial\Omega \tag{6}$$

$$\nabla \Phi_{MN} = U_{MN} \text{ on } \partial\Omega \tag{7}$$

Where $\sigma$ is the conductivity (inverse of the resistivity), $\Phi$ the potential distribution, $j$ the current density, $U$ the measured potential difference, $\Omega$ the domain and $\partial\Omega$ the edge of the domain. The calculations were carried out with the software package RES2DINV, which is developed for geophysical applications (Loke 2011).

## 4 Results

The influence of the electrode configuration on the reconstruction of the resistivity distribution in reinforced concrete will exemplary be shown by comparing the Wenner Alpha with the Dipole-Dipole configuration. The measurements were carried out with an electrode distance of 20 mm and a frequency of 25 Hz. The concrete specimens with dimensions of $400 \times 400 \times 200$ mm were stored in tap water until the measurement, so that the concrete is water saturated. A reinforcement bar with a diameter of 12 mm is centrically embedded with a concrete cover depth of 10 mm. The electrode array was positioned perpendicular to the reinforcement.

The determined resistivity distributions are shown in Fig. 6. The main focus lies on a good reproduction of the resistivity distribution in the concrete. Therefore it is important to reproduce the position of the reinforcement as good as possible to be sure to consider the influence of the reinforcement on the measured results adequately. It can be seen that the geometry of the reinforcement cannot be reproduced by using the Wenner configuration (Fig. 6a), whereas the Dipole-Dipole configuration (Fig. 6b) leads to a better approximation. This can be explained by comparing the sensitivity distributions of both configurations shown in Fig. 4. Due to the focused area of high and positive sensitivities of the Dipole-Dipole configuration it

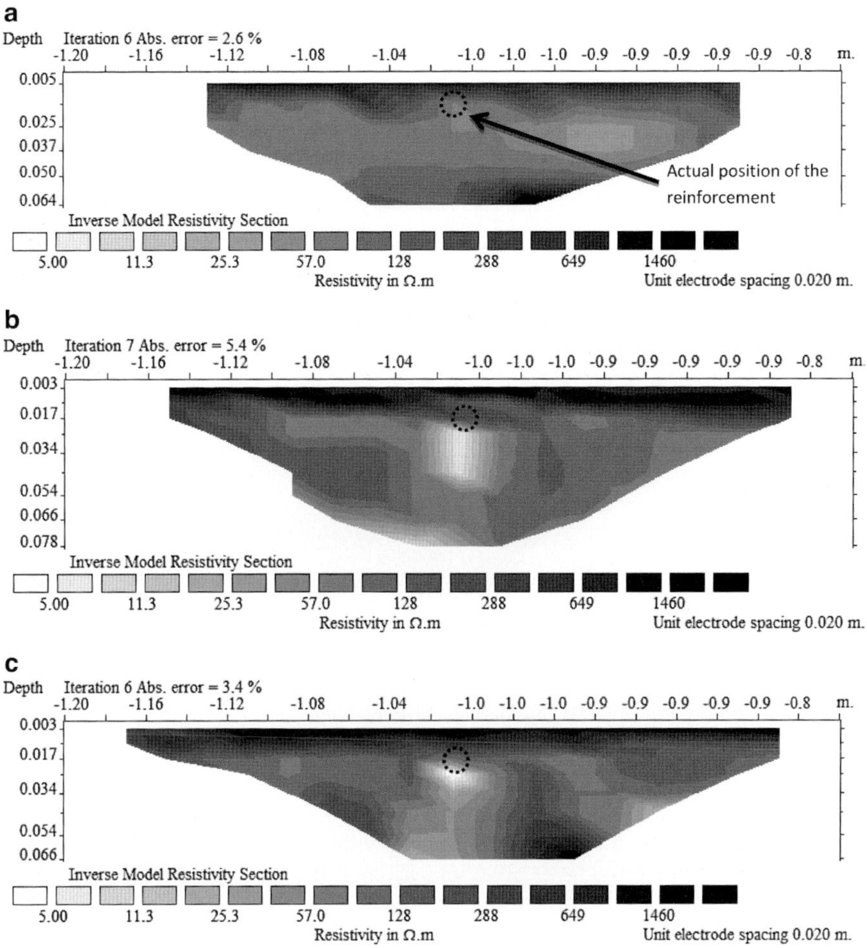


**Fig. 6** Resistivity distributions based on the Wenner Alpha (**a**), the dipole-dipole (**b**) configuration and a combined reconstruction (**c**); electrode distance 20 mm

is evident that by moving the electrode setup laterally the geometry of the reinforcement can be detected with a higher accuracy compared to the Wenner setup. The highest accuracy could be achieved by considering the information from both configurations (Fig. 6c).

Furthermore a resistivity variation near the surface can be noticed. By additional studies using NMR (Nuclear Magnetic Resonance, Antons and Raupach 2012) a higher porosity could be determined in this area. Due to the different grain size distribution at concrete surfaces the amount of hydrated cement is higher causing a higher porosity (Stoeckl 1966). In the case of water saturated concrete the increase in porosity correlates with a decrease in resistivity. The thin conductive layer

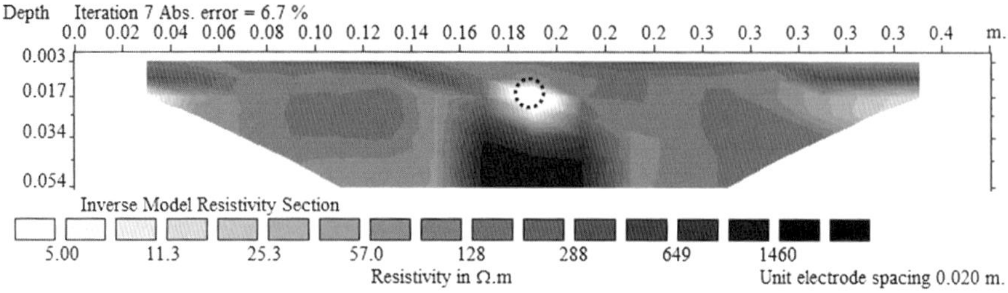


**Fig. 7** Resistivity distribution after removing a layer of 2 mm at the concrete surface, reconstruction under consideration of the dipole-dipole and Wenner Alpha configuration

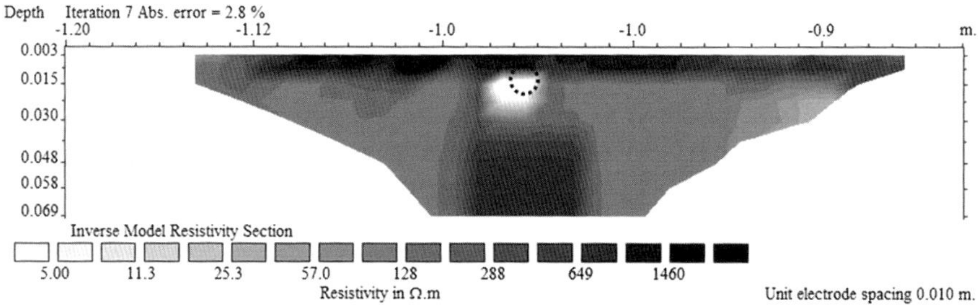


**Fig. 8** Resistivity distribution by considering a layer of 2 mm (10 Ωm) at the concrete surface, as well as the dipole-dipole and Wenner Alpha configuration

cannot be resolved by the ERT measurement with an electrode distance of 20 mm. Therefore the conductive layer gives rise to large negative sensitivities which are confirmed by numerical simulations. Low resistivity values and negative sensitivities lead to the mentioned paradoxal effect of high apparent resistivity values. In order to demonstrate this effect a layer of 2 mm was removed from the concrete surface and the measurements were repeated. The resistivity distribution is shown in Fig. 7.

It can be seen that the surface near resistivity profile could successfully be eliminated, which confirms the assumption mentioned above. In order to avoid the removal of the concrete on site the layer can be considered within the reconstruction of the resistivity distribution. In Fig. 8 the resistivity distribution is shown based on the same measurement data from Fig. 6c but considering a layer with a height of 2 mm and a resistivity of 10 Ωm. It can be seen that the resistivity distribution is more homogeneous compared to Fig. 6c and the depth of the reinforcement correlates with the actual concrete cover depth which leads to a rise in the accuracy of the reconstructed image. It can also be seen that the area under the reinforcement shows relatively high resistivity values. This is based on the low current densities during measurement which lead to low sensitivity values which can result in such artefacts.

# 5 Conclusions

Electrical resistivity tomography is a promising method which allows a precise and reliable determination of the resistivity distribution in the concrete. Not a single integral resistivity value but the apparent resistivity distribution is of interest. By reproducing the reinforcement its influence is considered to obtain a realistic distribution of the concrete resistivity. The interpretation of the results regarding calculation artefacts can be complicated why the mathematics must be understood by the operator.

Knowing the resistivity distribution allows a determination of the distribution of inhomogeneous material properties. This will allow spatial assessment of e.g. water saturation degrees, chloride diffusion coefficients, corrosion rates or porosities.

The studies will be continued e.g. by applying ERT on real structures with rebar meshes and moisture differences.

**Acknowledgements** Funded by the Excellence Initiative of the German federal and state governments. The measurements were carried out in cooperation with Prof. Christoph Clauser and Dr. Norbert Klitzsch from the "Institute for Applied Geophysics and Geothermal Energy" of RWTH Aachen University. The authors would like to thank Udo Antons for the helpful support regarding the NMR measurements.

# References

Andrade C, D'Andrea R, Castillo A, Castellote M (2009) The use of electrical resistivity as NDT method for the specification of the durability of reinforced concrete. Paper presented at the 7th international symposium on non destructive testing in civil engineering (LCPC 2009), Nantes, 30 June–3 July 2009

Antons U, Raupach M (2012) Zerstörungsfreie Visualisierung von hydrophob wirksamen Schichten in Beton. Restor Build Monum 18(5):287–302

Barker RD (1979) Signal contribution sections and their use in resistivity studies. Geophys J R Astron Soc 59(1):123–129. doi:10.1111/j.1365-246X.1979.tb02555.x

Castellote M, Andrade C (2006) Round-Robin test on methods for determining chloride transport parameters in concrete. Mater Struct 39(10):955–990. doi:10.1617/s11527-006-9193-x

Friedel S (2000) Über die Abbildungseigenschaften der geoelektrischen Impedanztomographie unter Berücksichtigung von endlicher Anzahl und endlicher Genauigkeit der Meßdaten, Shaker

Lines LR, Treitel S (1984) Tutorial – a review of least-squares inversion and its application to geophysical problems. Geophys Prospect 32(2):159–186

Loke MH (2011) RES2DINV v.3.5: Geotomo Software Sdn Bhd. Retrieved from www.geoelectrical.com

Polder R, Andrade C, Elsener B, Vennesland O, Gulikers J, Weidert R, Raupach M (2000) Rilem TC 154-EMC: electrochemical techniques for measuring metallic corrosion – test methods for on site measurement of resistivity of concrete. Mater Struct 33(234):603–611. doi:10.1007/Bf02480599

Rauen (2012) Geotest v 2.39: Geophysik Dr. Rauen. Retrieved from http://geophysik-dr-rauen.de

Raupach M (2006) Models for the propagation phase of reinforcement corrosion – an overview. Mater Corros (Werkstoffe und Korrosion) 57(8):605–613

Stoeckl S (1966) Das unterschiedliche Verformungsverhalten der Rand- und Kernzone von Beton. Deutscher Ausschuss für Stahlbeton (Heft 185), Berlin

Viergever MA, Todd-Pokroper AE (1988) Mathematics and computer science in medical imaging. Springer, New York

Warkus J, Raupach M (2010) Modelling of reinforcement corrosion – geometrical effects on macrocell corrosion. Mater Corros (Werkstoffe und Korrosion) 61(6):494–504

Wenner F (1915) A method of measuring earth resistivity. Bull Bur Stand 12:469–478

Weydert R, Gehlen C (1999) Electrolytic resistivity of cover concrete: relevance, measurement and interpretation. In: Proceedings of the durability of building materials and components 8, vols 1–4, pp 409–419. NRC Research Press, Ottawa ON KIA OR6, Canada. ISBN 0-660-17737-4

# Electronic and Electrolytic Conduction of Cement Pastes with Additions of Carbonaceous Materials

C. Andrade, P. Garcés, F.J. Baeza, Ó. Galao, and E. Zornoza

**Abstract** Additions to concrete may change some of its basic properties in several ways depending on the nature of these additions. The addition of fibers, in particular, has enabled new concrete characteristics, making standard concrete a very modern composite material. If the fibers are electrical conductors, the properties that change in addition to the mechanical ones, are the thermal and electrical conductivities. The results indicate that the arrangement of the electrodes and the electrode-material interface are relevant, because the use of sponges between electrode and concrete prevents the contact between the metallic electrode and the carbon material which ends in different values of electrical resistance with and without sponges. Moisture conditions, that critically influence the electrical resistance of concrete without additions, resulted, also very relevant when conductive substances are present in the matrix. If the proportion of the carbonaceous addition, that lowers significantly the resistivity, is to be quantified, the best procedure seems to measure in dry concrete (0 % relative humidity) with sponges or, alternatively, wet concrete (100 % relative humidity) with silver painted electrodes (without sponges).

**Keywords** Cement composites • Electronic conduction • Electrolytic conduction • Carbon fiber • Graphite powder • Conductivity

## 1 Introduction

Concrete is a very versatile material because its properties can be enhanced by the addition of other materials such as admixtures, minerals, fibers, etc., leading to a more modern composite material. These new mixes enable multiple applications of

C. Andrade (✉)
Department of Durability, Institute of Construction Science "Eduardo Torroja" (IETcc-CSIC), Madrid, Spain
e-mail: andrade@ietcc.csic.es

P. Garcés • F.J. Baeza • Ó. Galao • E. Zornoza
Department of Construction Engineering, University of Alicante, Apdo. Correos 99, 03080 Alicante, Spain
e-mail: pedro.garces@ua.es



C. Andrade et al. (eds.), *Durability of Reinforced Concrete from Composition to Protection*, DOI 10.1007/978-3-319-09921-7_2

concrete. The fibers, in particular, collaborate to increase its tensile strength and its ductility, for example. Various types of fibers have been used to accomplish these improvements, polypropylene, stainless steel and carbon fibers (ACI report #544.1R-96 2002). With respect to the latter as well as graphite powder (GP), their addition to a concrete mix will also influence its electrical resistance, thermal conductivity, ability to shield electrical fields, etc, which has resulted in concrete being considered as a multifunctional material (Chung 2003).

This paper is focused on the changes in electrical resistance due to the presence of carbon powder and fibers and the manner in which this resistance is measured. The electrical resistance of concrete was studied already in 1955 (Hammond and Robson 1955) as the electrical insulating properties of concrete were of interest to demonstrate its adequate performance in fabricating rail sleepers. It was well identified then that concrete behaves as a dielectric when it is almost completely dry and presents resistances in the range of a semiconductor when it is wet or water saturated. The conduction was identified to proceed through the migration of ions in the porous network (Monfore 1968). That is, it can behave as an electrolytic conductor. Further to the initial studies, resistivity levels in concrete have been identified to be closely related to setting occurrence, progression of hydration (Li et al. 2007) and to be the controlling parameter of reinforcement corrosion (Alonso et al. 1988) precisely due the resistivity is directly proportional to the degree of water saturation.

The addition of carbon fibers, while not contributing to the ionic concentration in the pore water, is expected to influence the electrical resistivity in other ways. Being the carbonaceous materials electronic conductors themselves, they may change the concrete total resistivity, particularly when the amount of addition surpasses the percolation threshold. This percolation threshold refers to the minimum quantity of carbonaceous material needed to create a continuous electronic conductive path. At this threshold the conductivity of the material experiments a significant decrease. This change may help to quantify the amount of addition present in the material. This phenomenon will occur with metallic fibers as well. Therefore, in the presence of these kinds of electrically conductive fibers there will be two types of electrical conduction in the concrete: the electrolytic one due to the ions in the pore network, and a here called "electronic" one due to the carbonaceous additions. The object of this article is to study the differences in the electrical resistance introduced by the addition of different proportions of these carbonaceous materials, to mortar samples exposed to various levels of humidity and to find the best experimental procedure to identify the percolation threshold, or at least the starting of the percolation phenomenon, due to the presence of an increasing amount of electrical conductive material.

## 2 Experimental Procedure

### 2.1 Materials and Specimen Preparation

Portland cement type CEM I 52.5 R (OPC) according to European standard (UNE-EN 2000) was used, and mixed with distilled water in all instances. Prismatic

**Table 1** Chemical composition of carbon additions

| | C | H | N | $(S+O)_{dif}$ |
|---|---|---|---|---|
| GP | 99.00 % | – | – | 1.00 % |
| SCF | 87.46 % | 0.68 % | 1.04 % | 10.57 % |
| LCF | 92.91 % | 0.18 % | 5.15 % | 1.76 % |

cement paste samples of $40 \times 40 \times 160$ mm dimension were prepared with a 0.5 water-cement ratio. Carbonaceous materials were added in the following amounts with respect to cement mass: 0 % (control), 0.5 %, 1 %, 2 %, 5 % of graphite powder (GP), short carbon fibers (SCF) and long carbon fibers (LCF). Besides, partial cement substitution of 25 and 50 % of graphite powder was used.

Commercially available coal tar pitch milled carbon fibers (SCF) provided by Donacarbo, were used for this study (general purpose carbon fiber type). The diameter and the length of this fiber are 13 μm and 130 μm, respectively. In addition, PAN based carbon fiber (LCF) provided by Hexcell was used. The diameter and the length of this fiber are 7.1 μm and 10 mm, respectively. Table 1 contains the chemical composition of carbon additions: GP, SCF and LCF.

The samples were stored in different environments corresponding to relative humidities (RH) of 0, 60, 80, and 100 % which were provided by salt solutions placed in the bottom of the chambers until constant weight of samples was achieved. The 0 % RH was obtained by drying in an oven at 105 °C until constant weight, and using silica gel in the bottom of the desiccator instead of a conditioning salt solution.

## 2.2 *Techniques*

To measure the electrical resistance, a distinction is made on whether or not sponges are introduced between the electrodes and the cement paste. The sponges prevent the direct contact of the metallic electrodes to the carbonaceous materials and then this arrangement (with sponges) is expected to provide the electrolytic resistance of the paste modified by the fact that part of the matrix is substituted with an electrically conductive material that may or not be percolated in function of the proportion added, (an induction phenomenon in the carbon materials will also occur that is not relevant for present research). Then, in the disposition "with sponges", two identically-dimensioned $4 \times 4$ cm copper plates were placed at either end of the sample (Fig. 1 down). Between each plate and the sample, wet sponges moistened with tap water were introduced to ensure electronic insulation of the metallic electrodes with the carbonaceous materials. The contact pressure between metallic electrodes-sponges-cement pastes was controlled applying a constant load of 100 N between the plates.

In the disposition "without sponges" the electrodes will be in direct contact to the electrically conductive material and then, the resistance will be much lower than with sponges and will depend on whether or not the electrical conductive material is

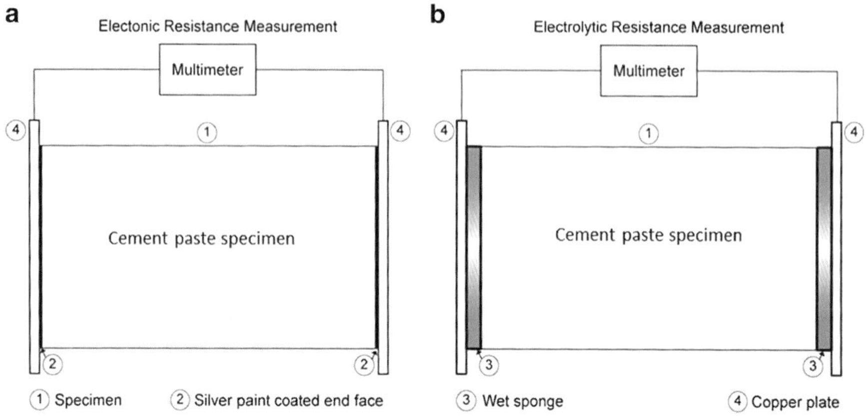


**Fig. 1** Measurement arrangement "without sponges" (**a**: electronic resistance measurement) and "with sponges" (**b**: electrolytic resistance measurement) in order to measure the electrical resistance

percolated, depending on the proportion by volume added. To measure the resistance (Fig. 1 up), the two end faces of the electrodes were painted with silver conductive paint to ensure good contact between the paste having the carbonaceous materials and the mains of the resistivimeter.

Resistance measurements were made with a Keithley 2002 multimeter. To calculate the resistance, Ohm's law was used ($R = V/I$) and $\rho = RA/L$, where $R$ = measured electrical resistance, $V$ = applied voltage, $I$ = current circulating due to the applied voltage, $\rho$ = resistivity, $A$ = cross section area (16 $cm^2$), $L$ = length of the specimen (16 cm).

# 3 Results

## *3.1 Measurement of the Electrical Resistance*

Figure 2 up shows the electrical resistance in the arrangement with sponges corresponding to cement samples with addition of GP in proportions of 0.5, 1, 2, and 5 %, and substitution in proportions of 25 and 50 %, in environments of RH of 0, 60, 80, and 100 %.

As may be observed in Fig. 2 up, the measured electrolytic resistance is much higher with sponges than without them and depends on the RH, as expected. At RH of 100 %, the resistance with sponges is very low, between 4.3 and 0.6 kΩ, for the control sample and substitution of 50 % GP, respectively. For the RH of 60 and 80 %, the resistance is higher but very similar to that at 100 % RH. The specimen at 80 % RH shows a slight, but negligible, increase in the resistivity with the smallest additions of GP. Apart from that, which may be within the measurement error,

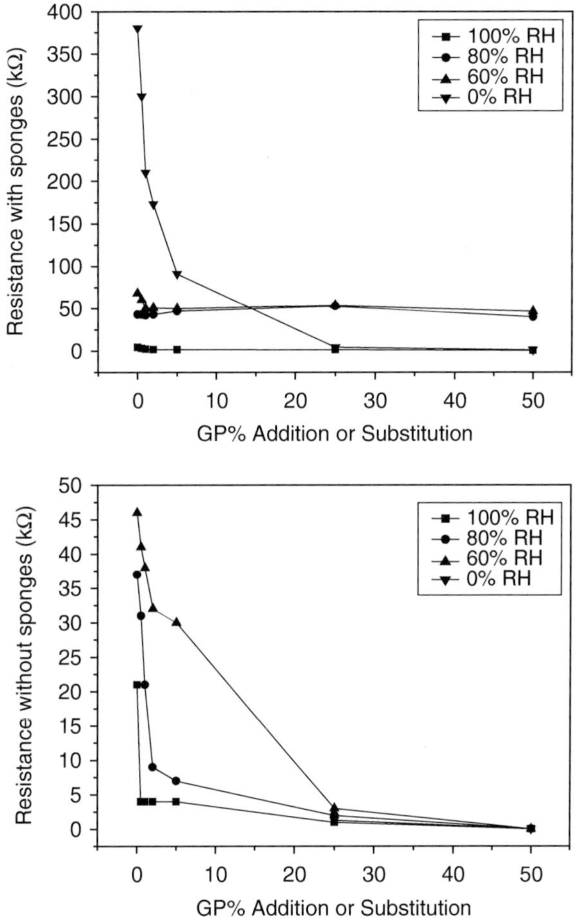


**Fig. 2** Electrical resistance measurements with sponges (*up*). Electrical resistance measurements without sponges (*down*). Cement samples with addition of graphite powder in proportions of 0.5, 1, 2, and 5 %, and cement substitution in proportions of 25 and 50 % at different RH. All carbon material percentages are referred to cement mass

similar to the case of 100 % RH, it seems that the additions at 60 and 80 % RH do not influence the electrolytic resistance. For the dry specimen (0 % RH) however, the effect of additions of GP is very notorious, with dramatically decreasing values of the electrolytic resistance with small additions and a very significant decrease for the 25 % substitution, reaching values similar to 100 % RH and smaller than 60 and 80 % RH. At this addition of 25 %, it seems that almost full percolation in electronic conduction has been reached.

Figure 2 down shows the values of the resistance without sponges corresponding to cement paste samples with GP, added also in increasing proportions of 0.5, 1, 2 and 5 % with respect to cement mass and substituted in proportions of 25 and 50 %. It is worth noting that the change in the scale of resistances due to their value is now much smaller. For the 0 % RH, results are plotted only for values >25 % of addition because those with smaller amounts are out of the chosen axis scale of resistance. For other RH values, in which the resistance significantly lowers, the decrease in resistance value is dramatic at the lower GP proportion and is gradual for higher amount of graphite powder. Then, without sponges in the low part of the figure, percolation seems to appear for very low amounts of carbonaceous powder at 100 % RH. The percolation initiation is detected at increasing amounts when the RH lowers, being at 25 % substitution at 60 % RH. That is, each disposition of electrodes shows different significant lowering of the resistance for different RH's and then, the percolation concept (decrease in resistance) needs to be defined for each arrangement and humidity level, as this last one can mask or influence significantly the detection of the decrease in electronic conduction.

Figure 3 up shows the resistance with sponges corresponding to cement paste samples with SCF additions (0.5, 1, 2 and 5 %), measured in the same environments of RH at 0, 60, 80 and 100 %.

In the arrangement with sponges (Fig. 3 up), the behavior with the SCF is very similar to that of the GP, but with a different proportion where dramatic resistance lowering is noticed. The specimens held at 60, 80 and 100 % almost do not exhibit

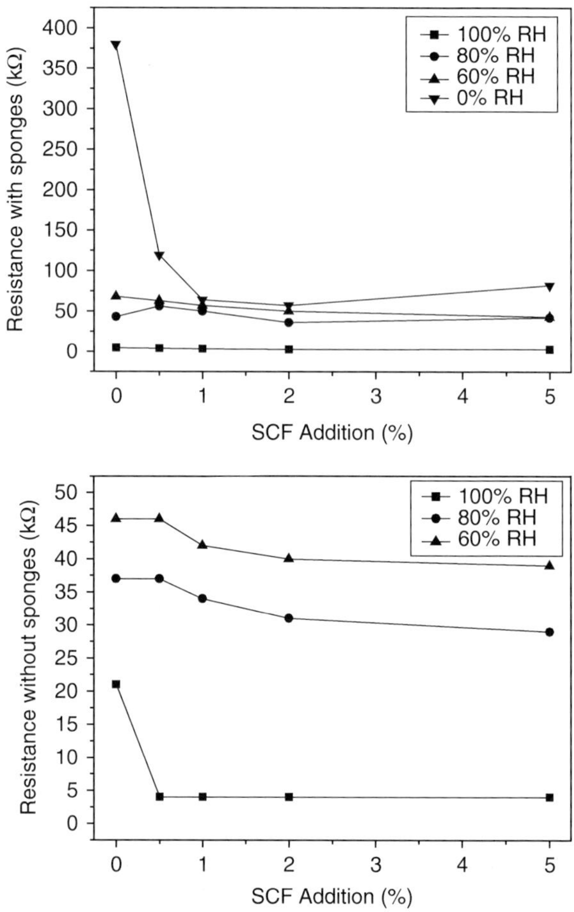


**Fig. 3** Electrical resistance measurements with sponges (*up*). Electrical resistance measurements without sponges (*down*). Cement samples with addition of short carbon fibers in increasing proportions of 0.5, 1, 2 and 5 % at different RH. All carbon fiber percentages are referred to cement mass

changes due to the additions, whereas the specimen in dry conditions (0 % RH) presents an indication of the percolation at 1 % of addition. Therefore, the presence of an electronic conductor influences the resistance measured with sponges only for very dry conditions, when there is almost no water in the capillary pores. At 60, 80 and 100 % RH, where water fills the small and large capillaries, the ionic conduction is more significant, masking the measurement of electronic conduction percolation.

The arrangement without sponges is shown in Fig. 3 down. Again the scale of resistances is very different. Specimens at 0 % RH were not tested due to the high values of resistance presented. In the case of 60 and 80 % RH, additions up to 5 % did not introduce significant decreases in the resistance. However, for 100 % RH, the effect of the addition is relevant from the smallest proportion of 0.5 % and the initiation of the percolation can be noticed from 0.5 % addition.

Figure 4 shows the electrical resistance measurements with and without sponges of cement pastes with addition of LCF, 1 cm length, in increasing proportions of

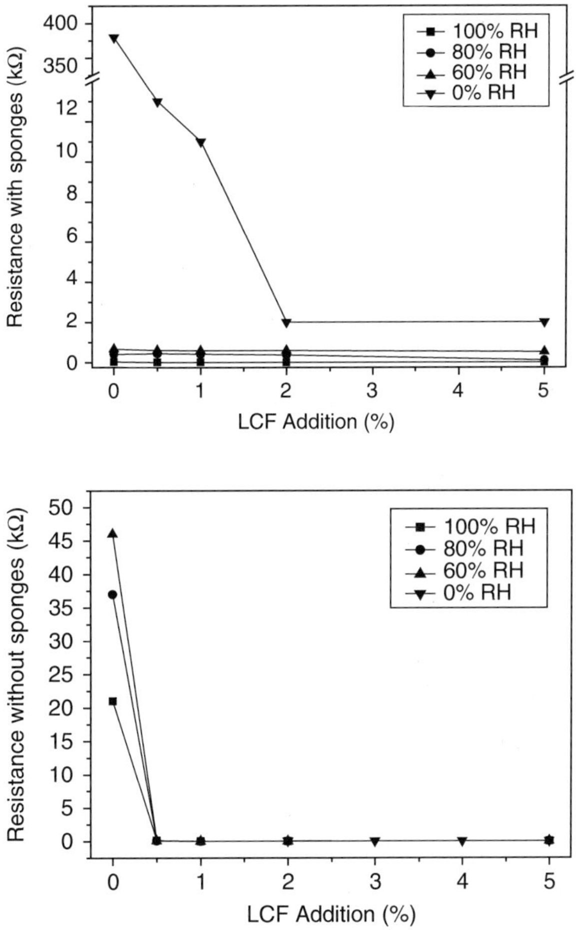


**Fig. 4** Electrical resistance measurements with sponges (*up*). Electrical resistance measurements without sponges (*down*). Cement samples with addition of the long carbon fibers in increasing proportions of 0.5, 1, 2 and 5 % at different RH. All carbon fiber percentages are referred to cement mass

0.5, 1, 2 and 5 % at RH of 0, 60, 80 and 100 %. In Fig. 4 up, the data again show that only in the dry specimen the additions or substitution has a relevant effect and that the initiation of the percolation is achieved for additions of only 0.5 % of the long fibers. Figure 4 down shows the data on the electrical resistance without sponges. In this case, the effect of the presence of long fibers, which in theory have greater ability to percolate at lower proportions of additions, is clearly detected by the important decrease recorded at 0.5 % of LCF addition for all the RH tested.

As a summary it can be said that for the case without sponges the electronic conduction percolation is better detected at 100 % RH while 0 % RH was the best condition for the measurement with sponges. Thus, not only the amount and type of addition are important factors in the measured value of the electrical resistance but the RH also plays an important role in the detection of the percolation threshold beyond which the resistance decreases to a minimum.

## 4 Discussion

To better observe the percolation phenomenon at the different RH, Fig. 5 presents the resistance values without sponges of specimens with the three carbonaceous materials at RH of 80 and 100 %, respectively. At 100 % RH (Fig. 5 up), the initiation of the electronic percolation is produced for all types of addition at 0.5 %, whereas for 80 % (Fig. 5 down), this addition indicates initiation of percolation only for the LCF, whose size and shape favor the electrical contacts. Then the humidity may mask the detection of electronic conduction in the case of the painted electrodes without sponges.

Figure 6 shows all the results of resistance without sponges for 0.5 and 2 % additions. This figure shows that the percolation initiation can be detected by a dramatic decrease in resistance values compared to control specimen, this effect is better shown at 100 % RH for all kinds of carbonaceous additions.

Finally, Fig. 7 shows a summary of the resistances measured with sponges in all the conditions tested. It reveals more comprehensibly that the effect of the proportion and shape of the addition is best detected in the 0 % RH specimen. According to this figure, only the dry specimen (0 % RH) enables good detection of the influence of the addition/substitution in the electronic conduction. Additionally, at 0 % RH, the resistance drastically decreases as the length of the fibers increase. However, when capillary water is starting to be present (HR around and higher than 60 %), the ionic conduction through the pore network masks the conduction through the carbonaceous materials and the percolation initiation is not well observed.

### *4.1 Types of Measurement and Conduction Mechanisms*

As mentioned, the two electrode arrangement, with and without sponges, have led to very different values of the resistance, being much lower when no sponges are

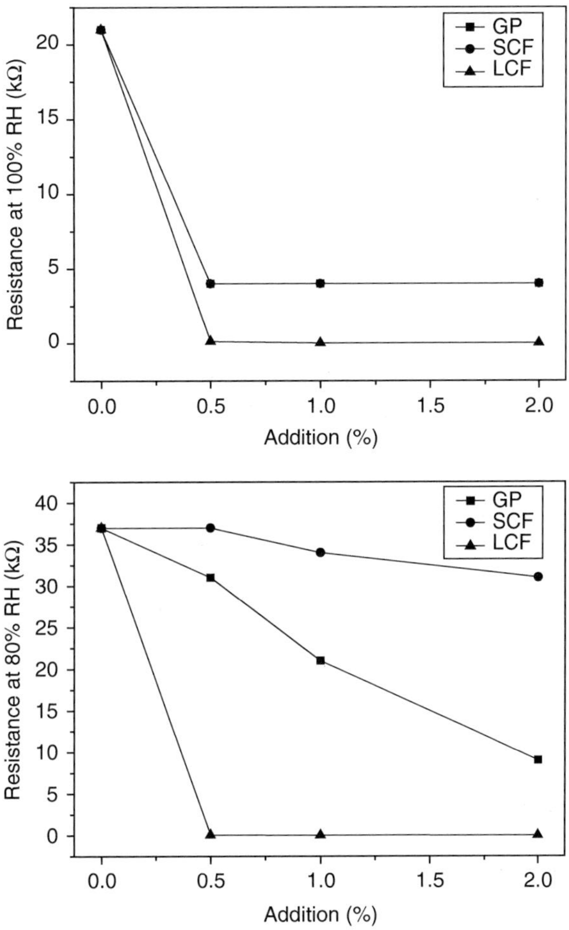


**Fig. 5** Resistance without sponges of specimens with the addition of three carbonaceous materials held in a chamber at 100 % RH (*up*) and 80 % RH (*down*)

used, that is when the electrodes can touch directly the electrical conductive material. This fact indicates the existence of two types of conduction mechanisms. When electrical conductive fibers are added to the mix, the resistance changes due to at least two factors: (1) the type of interfacial contact between electrodes and concrete and (2) the decrease of volume fraction of the cement phase (cement gel and capillary porosity) and the increase of that of the electrical conductor (the carbonaceous material).

When the electrodes are in direct contact with the conductive fibers or powder (as shown in Fig. 8), they will modify the measured value of the electrical resistance while they will not if sponges are placed between electrodes and cement paste. Sponges are electronic insulators that prevent the electronic conduction between the metallic electrodes and the electronic conductive materials.

Then, in the case of painted electrodes (without sponges), due to the interfacial contact of the electrodes and the carbonaceous material (Fig. 8), the measured

**Fig. 6** Resistance without sponges of specimens in different relative humidity atmospheres. Samples with 0.5 % of carbon material addition (*up*); samples with 2 % of carbon material addition (*down*)

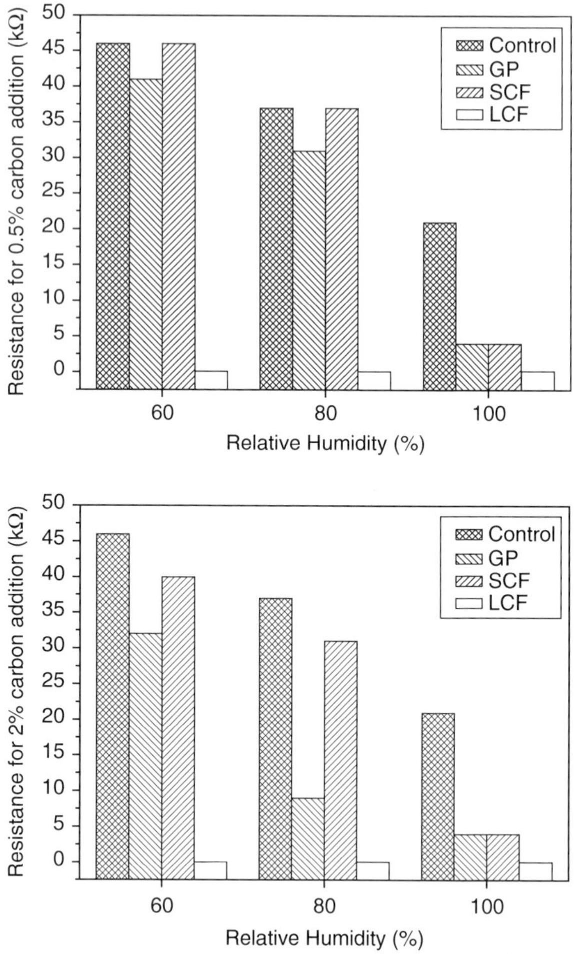


resistivity will be much lower than in the case with sponges were this contact electrode/carbon is not produced. Regarding the detection of the percolation due to the increase in the volume fraction of the carbonaceous material the better detection at the highest RH can be attributed to the fact that at in drier concrete the amount of conductive addition to compensate the higher resistances needs to be higher. That is the higher resistances mask the effect of the conduction of the carbon material.

In cement paste without conductive fibers, the conduction of electrical charges is made by the ions present in the pore solution, mainly that of capillary pores (Fig. 9). If the material is dry, then there is not ionic conduction and any decrease in resistivity due to the addition of fibers can be detected without any masking.

This lowering in resistance, although is not the main aim of present study to clarify it, is attributed to an electrostatic polarization, the same produced when measuring the resistivity by the four points method which recommends to avoid the

**Fig. 7** Resistance with sponges of specimens in different relative humidity atmospheres: Samples with 0.5 % of carbon material addition (*up*); samples with 2 % of carbon material addition (*down*)

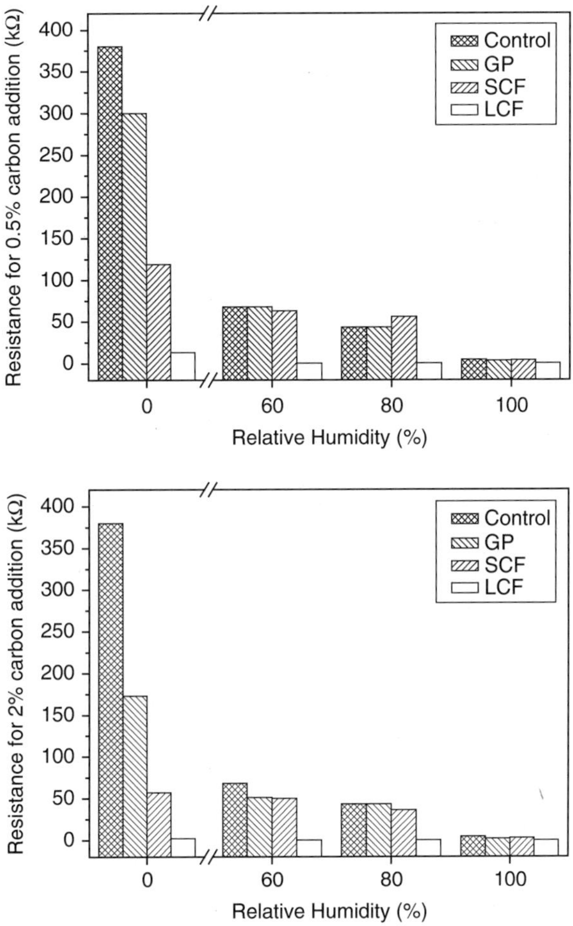


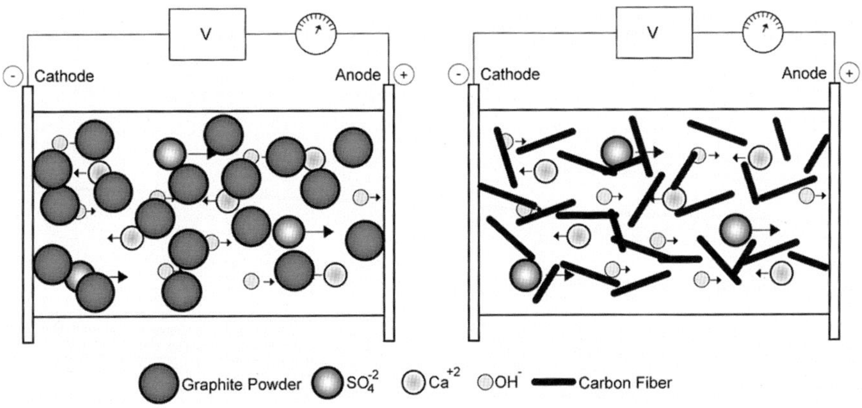


**Fig. 8** Representation of electrolytic and electronic conduction due to the direct contact of electrodes with the carbonaceous materials

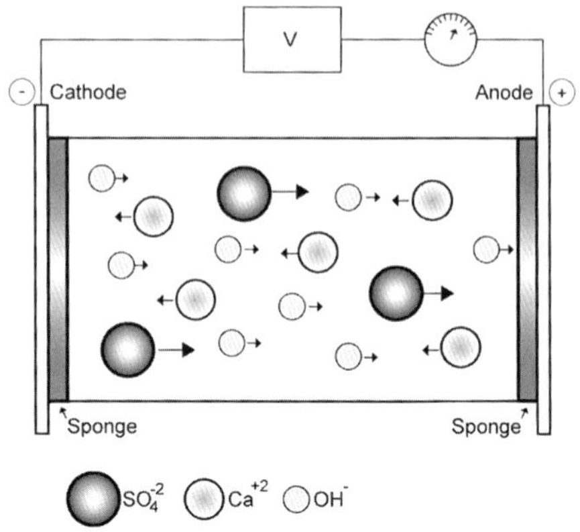


**Fig. 9** Representation of the movement of ions due to the application of a continuous current: electrolytic conduction

presence of rebars below. In this method, four electrodes are placed in the concrete surface to measure the resistivity and a lowering in the value is detected when placed just on reinforcements (Castellote et al. 1999, 2002; Feliú et al. 1996). The phenomenon of the lower resistivity values with respect when no bars are below is assumed to be the same than here makes possible to detect the addition of the fibers in a dry concrete.

## 4.2 Effect of the Amount of Addition: Percolation Initiation for Electrical Conduction

Percolation has been considered to initiate when beyond a value where almost no decrease in electrical resistance is observed. For smaller additions, there might be "cluster" or "aggregations" of electronic conductive paths (see Fig. 8), which explain the progressive lowering of measured values of resistance. In order to analyze the effect of the volume addition, Figs. 10 and 11 represent the resistance decrease with respect to the value with 0 % addition (100 %) and the volumetric fraction added. As mentioned, the lowest threshold amount is best detected at 0 % RH for resistance measurements with sponges and at 100 % RH for measurements without sponges (painted electrodes).

## 4.3 Effect of the RH in the Measurements

In spite of smallest amount of carbonaceous material is detected at the lowest and highest humidities depending upon the electrode arrangement, it may be of interest

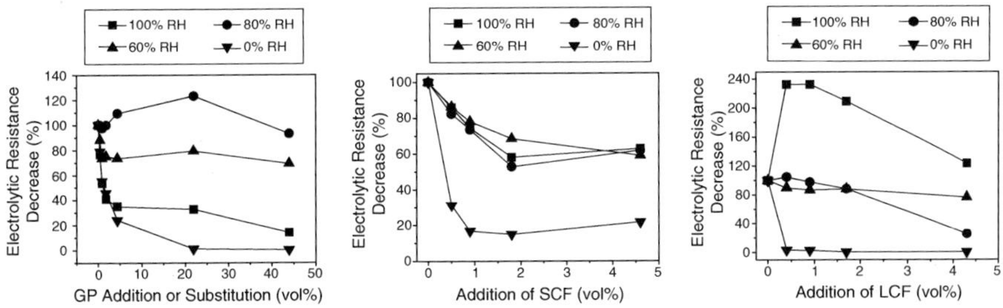


**Fig. 10** Dependence of the relative decrease (% with respect to the value with 0 % addition) of measurements with sponges in function of the volume of addition of GP (*left*), of SCF (*middle*) and of LCF (*right*)

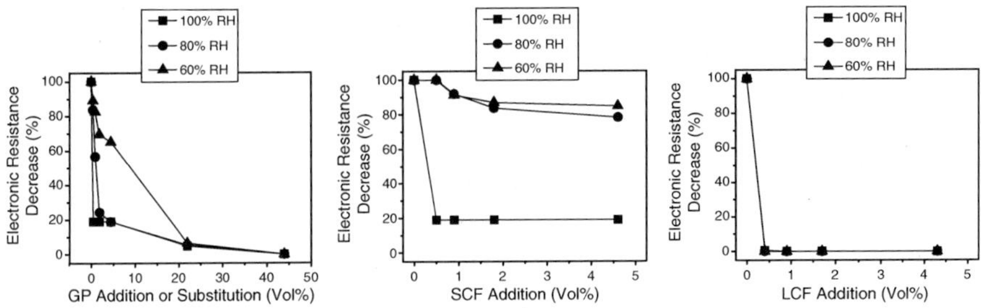


**Fig. 11** Dependence of the relative decrease (% with respect to the value with 0 % addition) of measurements without sponges in function of the volume of addition of GP (*left*), of SCF (*middle*) and of LCF (*right*)

to know if the addition of these materials significantly affects the values of the resistances at the normal humidities found in concrete structures. Present tests have shown the effect of moisture on the resistance values and how the percolation threshold varies with the moisture conditions (Figs. 10 and 11) and therefore, the humidity has to be specified when defining a threshold amount (Wen and Chung 2006; Ozyurt et al. 2006; Cheng et al. 2004).

## 4.4 Effect of the Fiber Aspect Ratio the Measurements

There is a significant difference in the levels of resistance of the samples containing GP, SCF and LCF. This can be interpreted as due to the difference in the size and shape of the carbonaceous materials, because the longer length of the fibers permits more continuous conduction paths (see Fig. 8), leading to the greater probability of fiber-fiber contact along the sample length. Therefore, as a general trend, lower amounts leading to percolation can be expected for longitudinal than for spherical particles, as well as larger ones.

# 5 Conclusions

The following conclusions can be drawn:

1. The electrical resistance of a cement-based material depends on the: (a) the volumetric fraction of solid cement phases (acting as an insulator) and porosity (acting as ionic conductor if water is present) and (b) the degree of water saturation in the pores equivalent to RH if equilibrium exists in controlled chambers.
2. If electronic conductive materials are added in the mix, the electrical resistance decreases in a proportion depending on the: (a) the type of carbonaceous material, (b) the size and shape of the material (longitudinal fibers induce a lower percolation threshold) and (c) the amount added (volumetric fraction in the mix)
3. Electrolytic and electronic conductions are produced beyond noticing the lowering of the resistance. For higher amounts of addition than those of the percolation initiation, the electronic conduction prevails and then the resistance decreases dramatically.
4. The interfacial contact between the material and the electrodes is very relevant if conductive materials are added, because the traditional use of a sponge between the electrodes and the specimen surface does not allow accounting for the complete electronic conduction.
5. For detecting the minimum threshold amount of conductive material the best is to measure with sponges in fully dry specimens (0 % RH). Alternatively if the electrodes are painted in the surface, the specimens should be conditioned at 100 % RH.
6. The proportion of conductive addition has different effects, depending not only on the shape but also on the humidity content of concrete. Therefore, care should be taken to identify the moisture content of the specimens when trying to account for the amount of fibers added by comparing the resistances of mixes with conductive materials, because the different degrees of saturation modify the absolute scale of resistance value.

**Acknowledgements** The authors would like to acknowledge financial support received from the Generalitat Valenciana (Spain) (CTIDIB/2002/164), from Ministerio de Ciencia y Tecnología (MAT 2003-06863) (BIA2006-10703), from Ministerio de Ciencia e Innovación MAT2009-10866 (subprograma MAT). E. Zornoza also thanks to Ministerio de Ciencia y Tecnología (Spain) for its post-doctoral support by Juan de la Cierva Programme.

# References

ACI report #544.1R-96. Report on fiber reinforced concrete (reapproved 2002). ACI Committee 544

Alonso C, Andrade C, Gonzalez JA (1988) Relation between resistivity and corrosion rate of reinforcements in carbonated mortar made with several cement types. Cem Concr Res 18:687–698

Castellote M, Andrade C, Alonso C (1999) Modelling of the processes during steady-state migration tests: quantification of transference numbers. Mater Struct 32:180–186

Castellote M, Andrade C, Alonso C (2002) Accelerated simultaneous determination of the chloride threshold and of the non-stationary diffusion coefficient values. Corros Sci 44:2409–2424

Cheng B, Wu K, Yao W (2004) Conductivity of carbon fiber reinforced cement-based composites. Cem Concr Compos 26:291–297

Chung DDL (2003) Multifunctional cement-based materials. Michael Meyer & Marcel Decker, New York

Feliú S, Andrade C, González JA, Alonso C (1996) A new method for in-situ measurements of electrical resistivity of reinforced concrete. Mater Struct 29:362–365

Hammond E, Robson TD (1955) Comparison of electrical properties of various cements and concretes. Engineer 199(5165):78–80

Li ZJ, Xiao LZ, Wie XS (2007) Determination of concrete setting time using electrical resistivity measurements. J Mater Civil Eng 19(5):423–427

Monfore GE (1968) The electrical resistivity of concrete. J PCA Res Dev Lab 10:35–48

Ozyurt N, Mason TO, Shah SP (2006) Non-destructive monitoring of fiber orientation using AC-IS: an industrial scale application. Cem Concr Res 36:1653–1660

UNE-EN, Spanish-European Standard 197-1:2000. Cemento. Parte 1: Composición, especificaciones y criterios de conformidad de los cementos comunes

Wen S, Chung DDL (2006) The role of electronic and ionic conduction in the electrical conductivity of carbon fiber reinforced cement. Carbon 44:2130–2138

# Influence of Anolyte on Lithium Migration in Concrete

Lourdes M.S. Souza, Rob B. Polder, and Oğuzhan Çopuroğlu

**Abstract** Alkali-silica reaction (ASR) affects concrete structures worldwide. During this deleterious process, alkali and hydroxyl ions react with reactive siliceous components of the aggregate, producing a hygroscopic gel. Once the gel absorbs water from the surrounding cement paste, it swells. Consequently, the reaction might lead to expansion and cracking of concrete elements. Lithium is known to prevent those detrimental effects. In fresh mixtures, the incorporation of lithium-based admixtures as a preventive method has been acknowledged for years. However, in hardened concrete lithium ions need to be driven into it. Ionic migration seems to be the most effective method, when compared to other transport mechanisms. Even though several investigations have been conducted on the use of electric field to transport lithium ions into concrete, so far, there is no agreement on the findings. The present paper aims to investigate the influence of the type and concentration of lithium solution used as anolyte.

**Keywords** Alkali-silica reaction • Repair • Migration • Lithium solutions

## 1 Introduction

Alkali-silica reaction (ASR) is a durability problem that affects concrete structures worldwide. During this reaction, alkali ions present in the pore solution react with certain reactive siliceous compounds present in the aggregate particle and, as a result, a hygroscopic alkali-silica gel is formed. The produced gel absorbs water from the surrounding cement paste and swells. This process might lead to deleterious expansion and further cracking of the concrete. In order to avoid ASR in new structures, there are already established procedures such as limiting the alkali

L.M.S. Souza (✉) • O. Çopuroğlu
Faculty of Civil Engineering and Geosciences, Materials and Environment,
Delft University of Technology, Delft, The Netherlands
e-mail: l.m.silvadesouza@tudelft.nl

R.B. Polder
Faculty of Civil Engineering and Geosciences, Materials and Environment,
Delft University of Technology, Delft, The Netherlands

TNO Built Environment and Geosciences, Civil Infrastructure, Delft, The Netherlands



C. Andrade et al. (eds.), *Durability of Reinforced Concrete from Composition to Protection*, DOI 10.1007/978-3-319-09921-7_3

content of the mixture or the use of supplementary materials. Nevertheless, once the reaction occurs in existing constructions, currently, there are no effective treatments.

Since 1951, it is known that the incorporation of lithium salts in mortars and concretes can reduce or even prevent ASR deleterious expansion (McCoy and Caldwell 1951). It is generally believed that lithium ions combine with the alkali-silica gel, forming a non-expansive gel (Feng et al. 2010).

In new concrete structures, lithium ions may be incorporated by the use of lithium based admixtures in the fresh mixture. In existing concrete structures, however, the ions need to be transported into the cementitious matrix. Amongst other transport mechanism such as diffusion and absorption, ionic migration seems to be the most effective method, leading to higher depths of penetration and concentration of lithium ions (Thomas et al. 2007; Santos Silva et al. 2008).

The use of an electric field to carry lithium ions inside concrete to treat ASR was first suggested by Page and Yu (1995). Since then, several studies have been carried out on the subject (e.g. Whitmore and Abbot 2000; Thomas and Stokes 2004; Santos Silva et al. 2008; Pacheco Farias and Polder 2010; Bentivegna et al. 2011; Liu et al. 2011; Ueda et al. 2013). However, there is no agreement on the conclusions. In fact, the variability of experimental conditions and procedures of the several studies makes it hard to compare and to draw conclusions. It is still necessary to fully understand some of the principles of lithium migration and the effect on ASR affected concrete. The present study intends to elucidate the influence of the solution used as anolyte on lithium migration, investigating different lithium compounds and solutions with varying concentrations.

# 2 Experimental

## *2.1 Materials and Specimen Preparation*

Mortar specimens were made with water to cement ratio (w/c) of 0.5 and sand to cement proportion of 3:1. They were cast in PVC cylinder molds of 98 mm of diameter and 50 mm of height. Mixing and compaction procedures followed EN 196 1:2005. After casting, the specimens were cured in a fog room for 36 days before testing. In the confection of the mortar, Portland cement type CEM I 42.5 N, standard sand with $D_{max}$ of 2.00 mm (according to EN 196 1:2005) and deionized water were used.

## *2.2 Apparatus and Procedure*

The two-chamber lithium migration test was performed in the set-up described by ASTM 1202. In this type of set-up, the specimen is placed between two chambers,

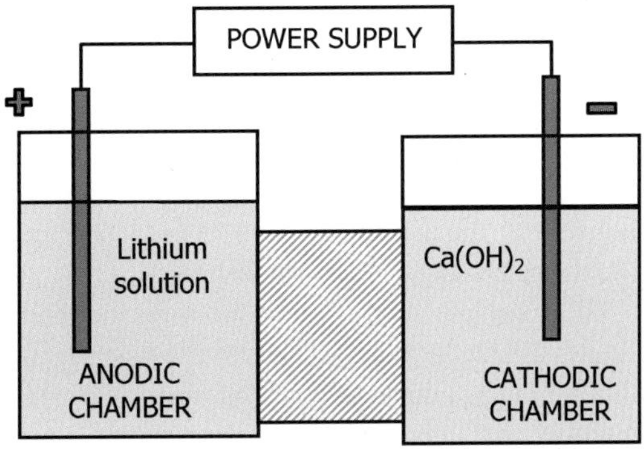


**Fig. 1** Schematic diagram of the experimental set-up

**Table 1** Lithium solutions used as anolytes

| Solution | | Concentration (M) | |
|---|---|---|---|
| $Li_2CO_3$ | 0.2 M (saturated) | | |
| LiOH | 0.2 M | 4.9 M (saturated) | |
| $LiNO_3$ | 0.2 M | 4.9 M | 7.8 M (saturated) |

each equipped with a metallic electrode, as shown in Fig. 1 In each chamber an electrolyte is poured. When an electric potential difference is applied between the electrodes, cations move towards the cathode, and the anions move in the opposite direction.

In each test, the anolyte was the investigated lithium compound solution, according to Table 1. The catholyte on the other hand, was a saturated $Ca(OH)_2$ solution in all tests. For each anolyte solution, two tests were performed. The applied voltage was 40 V and the tests ran for 1 week. During that time, passing current and catholyte temperature were continuously monitored. Electrical resistance and electrolyte pH were measured four times during the test. For the measurement of the electrical resistance of the specimen, a multimeter was used, at 120 Hz in the resistance mode. From the resistance, the resistivity was calculated by Eq. 1 below:

$$\rho = {}^{R*A}/_{L} \tag{1}$$

where $R$ is the measured electrical resistance (Ω), $A$ is the specimen surface area ($m^2$) and $L$ is the thickness of the specimen (m).

The pH measurements were made with a pH-meter, when it was possible. When the solution was very alkaline ($pH > 13$) or presented an elevated concentration of lithium ions, pH strips were used. Samples of electrolytes were collected three times during the experiment and were analyzed by Inductively Coupled Plasma (ICP), in order to obtain the concentration of sodium, potassium and lithium and calcium (the latter only in the anolyte).

# 3 Results

Figure 2a shows the current density behavior of all cells. It is necessary to highlight that electrical connection problems occurred in the cell of the test $LiNO_3$ – 4.9 M (2), as can be observed in the current density plot. In all cells, a rapid increase can be noticed in the first couple of hours. The same was observed by other authors (Pacheco Farias and Polder 2010; Liu et al. 2011) and it is believed to be related to the complete saturation of the mortar specimen with water (Pacheco Farias and Polder 2010). In most cells, after the first day, the current density drops until a value that remains somewhat constant until the end of the experiment. This behavior was not observed in both cells of $LiNO_3$ with concentration of 7.8 M and in cell $LiNO_3$ – 4.9 M (1). In any case, these results contradict the findings of Pacheco Farias and Polder (2010), where the authors observed that the current had decreased to values lower than 5 $A/m^2$ by the end of the test.

Figure 2b shows the temperature behavior of the cells. The temperature rose in all cells, what influenced the resistivity values, as will be discussed in more detail later. In most cases, it behaved similarly to the current density with a rapid increase in the first couple of hours, slower decrease after 1 day of testing to an approximately constant value until the seventh day.

The passing electrical charges, obtained from the integration of the current plots, are shown in Table 2. The presented values are the average between the charges of

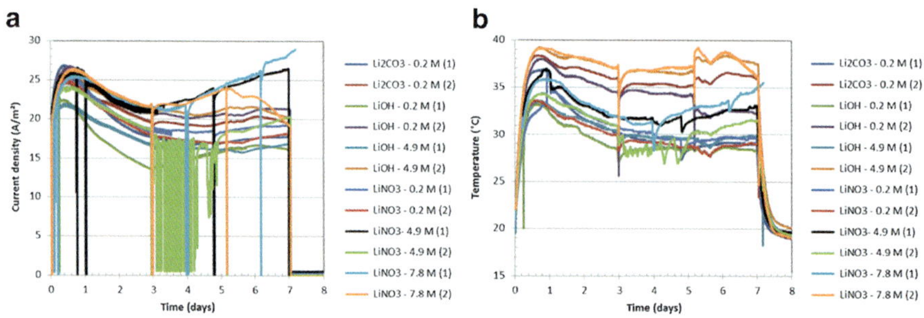


**Fig. 2** Current density (**a**) and temperature plots (**b**)

**Table 2** The passing charge of tested solutions

| Solutions | Charge (C) | CV (%) | Mean charge of solutions of same concentration (C) | CV of solutions of same concentration (%) |
|---|---|---|---|---|
| $Li_2CO_3$ 0.2 M | 94,758.25 | 3.80 | 92,921.00 | |
| LiOH 0.2 M | 91,361.05 | 14.29 | | 1.51 |
| $LiNO_3$ 0.2 M | 92,643.70 | 2.07 | | |
| LiOH 4.9 M | 93,442.85 | 10.87 | 96,750.58 | 3.42 |
| $LiNO_3$ 4.9 M | 100,058.30 | 11.84 | | |
| $LiNO_3$ 7.8 M | 112,434.50 | 3.16 | – | – |

*CV* coefficient of variation

the duplicates. The high variation of the solution $LiNO_3$ 4.9 M (11.84 %) can be explained by connection problems. Nevertheless, the same cannot be said of the solutions LiOH with concentrations 0.2 and 4.9 M. In these cases, no connections issues were detected and, as the tests were performed simultaneously, there were no differences in room temperature between duplicates. Further tests might explain those variations. It seems like that the passing charge increases with the concentration of the solution, although not too much. Also, when comparing solutions with same concentration, the values of passing charge did not vary by more than 5 %. This indicates that the type of anolyte solution does not significantly influence the charge passed.

The electrical resistivity variations and the temperatures of the cell when the measurements were made are shown in Fig. 3. Apart from the cell $Li_2CO_3$ 0.2 M (2), all cells presented an initial resistivity around 45–50 Ω.m. Comparing both graphs, it is possible to notice that resistivity variations followed temperature changes: higher temperatures led to lower resistivity. In any case, resistivity variations were reversible, i.e. they came back to their original values after the temperature went back to its original value. These results did not agree with the ones by Pacheco Farias and Polder (2010), where increase of resistivity was observed from the start to the end of the test.

The pH progress during the experiment can be seen in Fig. 4. Due to problems with the pH-electrode, pH values were not measured in cells $Li_2CO_3$ 0.2 M (1) and

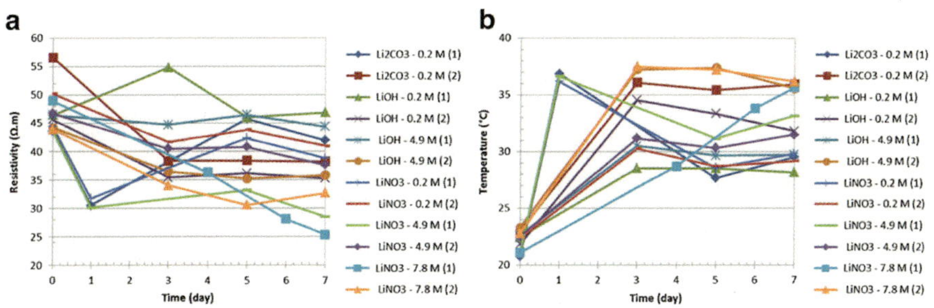


Fig. 3 Electrical resistivity (a) and temperature of measurement (b)

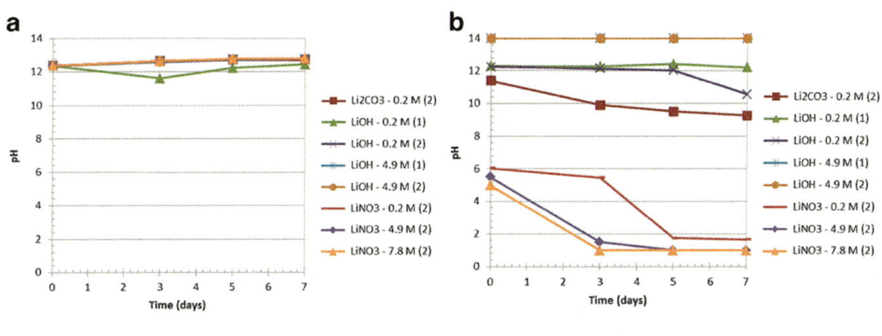


Fig. 4 pH of catholyte (a) and anolyte (b)

$LiNO_3$ 0.2 M, 4.9 M and 7.8 M (1). With the exception of LiOH – 0.2 M (1), all catholyte solutions exhibited a fairly stable value, around pH 12, with a slight increase during the test. The pH of the anolytes, on the other hand, depend on the initial pH, the passing charge and the buffer capacity of the solution. The $LiNO_3$ solutions initially had an acid pH and they showed the highest drops in pH. The LiOH – 4.9 M solutions, on the other hand, were very basic and, even after the experiment, still presented high pH values. As for the anolytes of $Li_2CO_3$ – 0.2 M (2) and LiOH – 0.2 M (1) and (2), even though they exhibited close initial pH values, the latter showed better buffer capacity.

The pH variations in catholyte and anolyte are related to the cathodic and anodic reactions, respectively, shown in Eq. 2 below:

$$\begin{aligned} &2H_2O + 2e^- \rightarrow 2OH^- + H_2 && \text{Cathodic reaction} \\ &2OH^- \rightarrow H_2O + \frac{1}{2}O_2(g) + 2e^- && \text{Anodic reaction (high pH)} \\ &H_2O \rightarrow 2H^+ + \frac{1}{2}O_2(g) + 2e^- && \text{Anodic reaction (low pH)} \end{aligned} \quad (2)$$

At the moment the erratic behavior of both cathodic and anodic pH values of the cell LiOH 0.2 M (1) cannot be explained satisfactorily and further experiments are needed. It is worth noting that the anolyte of the cells with $LiNO_3$ solutions became very acidic by the end of the experiment.

In Fig. 5a–f, the variations of chemical compositions of anolyte and catholyte can be seen. As for the presence of sodium, potassium and calcium in the anolyte Fig. 5a–c, this can be explained by diffusion from the specimen. Once again, LiOH 0.2 M (1) presented unexpected behavior and further analysis is necessary before drawing definite conclusions about this experiment. Even though no particular trend can be seen, it is interesting to notice that the $LiNO_3$ 7.8 M induced higher concentration of those ions in the anolyte by the end of the test. This is possibly due to the low pH developed during the test that might have led to the dissolution of the surface of the specimen.

The concentration of lithium in the anolyte solutions (Fig. 6) was expected to decrease throughout the experiment, as the ions entered the specimen. However, even in cases where lithium was detected in the catholyte, and, therefore, there was transport, the plots presented increment. The increase of temperature was first considered as an explanation, as it could lead to evaporation of water and increase the concentration. But, this behavior was not observed in all cells. Therefore further investigation is still needed.

The concentrations of sodium and potassium in the catholyte (Fig. 5e–f) developed as expected. Potassium ions leave the specimen faster than sodium because of higher ionic mobility. Nevertheless, once again, no clear trend can be seen. The same can be said about the concentrations of lithium in the catholyte. Although $LiNO_3$ 4.9 and 7.8 M presented higher values of lithium concentration in the catholyte, there is no clear correlation between initial concentration and type of anolyte and final lithium concentration in the catholyte.

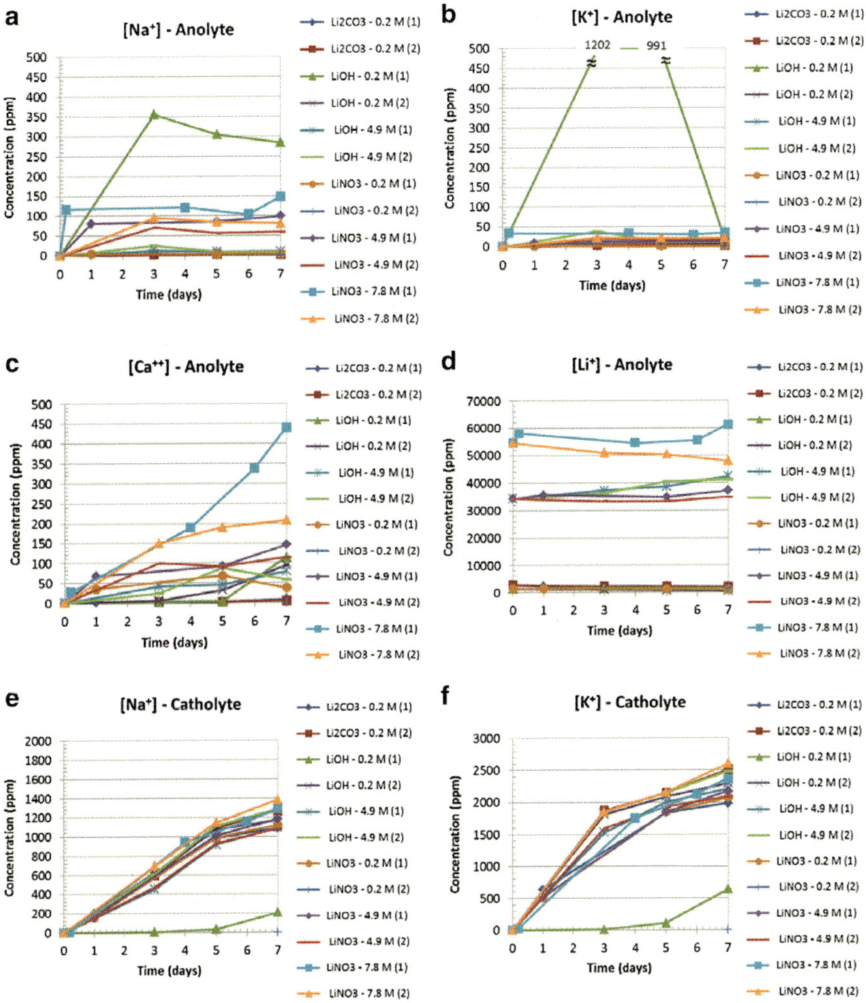


**Fig. 5** Concentration in anolyte solutions of sodium (**a**), potassium (**b**), calcium (**c**) and lithium (**d**) and concentration in catholyte solutions of sodium (**e**) and potassium (**f**)

**Fig. 6** Lithium concentration in catholyte solutions

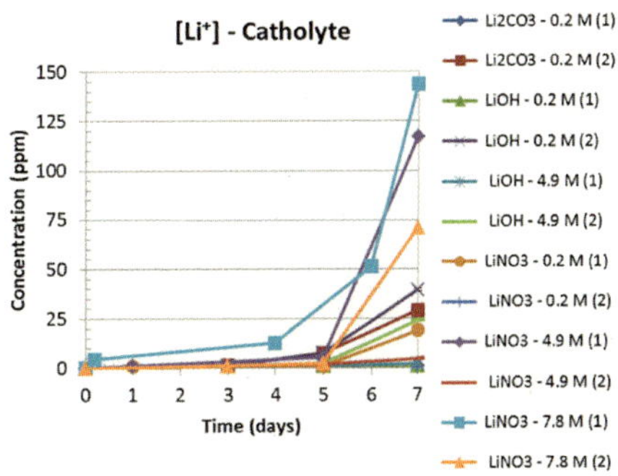

## 4 Concluding Remarks

From the presented study, it is observed that the lithium solution used as anolyte plays a role in the migration process. However, this is not yet very clear. Anolytes with higher concentration led to higher values of charge passing through the specimen. Nevertheless this behavior could not be directly related to the final concentrations of lithium, sodium, potassium and calcium in both chambers. Also, the initial characteristics of the anolyte influenced the pH development during the test. Finally, the resistivity variations during the experiment were attributed to temperature changes.

This study is a work in progress. Further experiments, such as obtaining the lithium profile of the specimens, are still necessary in order to better understand the mechanism behind the migration of lithium.

## References

Bentivegna AF, Giannini ER, Folliard KJ (2011) Use of electrochemical migration to mitigate alkali-silica reaction in large scale concrete structures. In Proceedings of the 4th international conference of concrete repair. Concrete Solutions, Dresden, Sept 2011

Feng X, Thomas MDA, Bremner BJ, Folliard KJ, Fournier B (2010) New observations on the mechanism of lithium nitrate against alkali silica reaction (ASR). Cem Concr Res 40(1):94–101

Liu C-C, Wang W-C, Lee C (2011) Behavior of cations in mortar under accelerated lithium migration technique controlled by a constant voltage. J Mar Sci Technol 19(1):26–34

McCoy EJ, Caldwell AG (1951) New approach to inhibiting alkali-aggregate expansion. J Am Concr Inst 22:693–706

Pacheco Farias J, Polder RB (2010) Preliminary study of electrochemical lithium migration into cementitious mortar. In: van Breugel K, Ye G, Yuan Y (eds) Proceedings of the 2nd international symposium on service life design for infrastructure, Delft, pp 1093–1100, Oct 2010. RILEM publications, Bagneux

Page CL, Yu SW (1995) Potential effects of electrochemical desalination of concrete on alkali-silica reaction. Mag Concr Res 47(170):23–31

Santos Silva A, Salta M, Melo Jorge ME, Rodrigues MP, Cristino AF (2008) Research on the suppression expansion due to ASR. Effect of coatings and lithium nitrate. In: Broekmans M, Wigum BJ (eds) Proceedings of the 13th international conference on alkali-aggregate reaction in concrete (ICAAR), Trondheim, June 2008

Thomas MDA, Stokes DB (2004) Lithium impregnation of ASR-affected concrete: preliminary studies. In: Mingshu T, Peking DM (eds) Proceedings of the 12th international conference on alkali-aggregate reaction in concrete, Beijing, pp 659–667, Oct 2004. Beijing World Publishing Corporation, Beijing

Thomas MDA, Fournier B, Folliard KJ, Ideker JH, Resendez Y (2007) The use of lithium to prevent or mitigate alkali-silica reaction in concrete pavements and structures. Report from Federal Highway Administration FHWA-HRT-06-133. U.S. Department of Transportation, Washington, DC

Ueda T, Baba Y, Nanasawa A (2013) Penetration of lithium into ASR-affected concrete due to electro-osmosis of lithium carbonate solution. Const Build Mater 39:113–118

Whitmore D, Abbot S (2000) Use of an applied electric field to drive lithium ions into alkali-silica reactive structures. In: Bérubé MA, Fournier B, Durand B (eds) Proceedings of the 11th international conference on alkali-aggregate reaction in concrete (ICAAR), Quebec City, pp 1089–1098, June 2000

# Durability of Hydrophobic Treatments on Concrete

**U. Antons, M. Raupach, and O. Weichold**

**Abstract** The paper deals with the durability of hydrophobic layers in concrete, investigated by using non-destructive single-sided nuclear magnetic resonance (NMR). With this measuring technology, the amount and distribution of water as well as of hydrophobing agents inside the concrete can be visualised. By comparing NMR measurements of single specimen over time, changes in the hydrophobic layer properties can be seen. Results after UV radiation, alkaline storage and carbonation show the different effects of each exposure. The investigated hydrophobing agents for application on concrete were stable during long term exposure in alkaline solution. The effect of UV radiation was limited to the roll-off behaviour of the concrete surface. The results after $CO_2$ exposure show that structural changes caused by carbonation of blast furnace slag cement based concrete decrease the hydrophobic layer thickness, while carbonation of Portland cement concrete has no influence. In addition, efficacy tests show that water transport through the affected hydrophobic layers of blast furnace slag concrete is accelerated.

**Keywords** Hydrophobing agent • Durability • Non-destructive testing

## 1 Introduction

Outdoor structures are exposed to a variety of chemical and physical stresses, and most of these are associated with the presence of water. Surface protection systems such as coatings, impregnations, or hydrophobing agents can be used to counteract these attacks. Due to the unchanged appearance of the structure after application, the use of hydrophobing agents is very attractive in many cases. Unfortunately the characteristic roll-off behaviour, which is commonly associated with hydrophobic surfaces, fades away after a certain period of exposure to ultraviolet radiation. Mainly due to this superficial reason, hydrophobing agents have a poor reputation in terms of durability aspects, although the hydrophobic layer below the affected surface was found to be still intact (Meier and Wittmann 2011; Polder et al. 2001).

U. Antons (✉) • M. Raupach • O. Weichold
Institute of Building Materials Research, RWTH Aachen University, Aachen, Germany
e-mail: antons@ibac.rwth-aachen.de



C. Andrade et al. (eds.), *Durability of Reinforced Concrete from Composition to Protection*, DOI 10.1007/978-3-319-09921-7_4

In this context, the German Research Foundation funded the research project "durability of hydrophobic treatments on concrete surfaces" with the aim to assess the extent to which certain exposures such as UV radiation, carbonation, hydration, and temperature affect the long term durability and efficacy of hydrophobing agents. As major indicators for the durability, the thickness and potential changes of the hydrophobic layer were evaluated. For this assessment, the so-called NMR-MOUSE® (Nuclear Magnetic Resonance – Mobile Universal Surface Explorer, Blümich 2004) was the method of choice. In addition to determining the hydrophobic layer thickness at different times, it also allows to follow the ingress of the hydrophobing agent as well as water.

Herein we present the effect of UV radiation, alkaline storage, and carbonation on the thickness and efficacy of hydrophobic layers. The parameters are determined by using the NMR-Mouse to locate the water in water saturated specimens with a hydrophobic treatment and to follow the transport of water through the hydrophobic layer.

## 2 Materials, Sample Preparation, and Measuring Method

Specimens based on two different types of cement (Portland cement, OPC (CEM I), and blast furnace slag cement, PBFC (CEM III/A, 65 % slag)) have been investigated. Both concretes had the same amount of aggregates (1,850 kg/m$^3$, grading curve A/B 16, DIN EN 12620), cement content (300 kg/m$^3$), and the same water cement ratio (0.6). The specimens were demoulded after 1 day and subsequently stored under water for 28 days at room temperature. During this storage the specimens were cut to their final disc shape of 100 by 30 mm. Before applying the hydrophobing agents, all specimens were conditioned at 23 °C and 50 % RH. One set of specimens was conditioned in an atmosphere containing 2 vol.-% of carbon dioxide until the carbonation front reached a depth of 10 mm, while a second set was conditioned for an equal period of time in the absence of carbon dioxide to prevent carbonation.

The presented results are limited to the hydrophobing agents listed in Table 1. These were applied only on the mould-plain bottom side of the specimens. The amount of active substance per square meter was kept constant causing the applied quantities to differ.

**Table 1** Type of active substance, applied quantity and purity of the hydrophobing agents

| Agent | Active substance | Applied quantity | Active substance content |
|---|---|---|---|
| – | – | g/m$^2$ | % |
| Hyd.-4 | *n*-Octyltriethoxy silane | 130 | 98 |
| Hyd.-5 | *n*-Butyltriethoxy silane | 130 | 98 |
| Hyd.-6 | Iso-Octyltriethoxy silane | 129 | 99 |
| Hyd.-7 | Alkoxy silane cream[a] | 159 | 80 |

[a]Water based cream

Due to the size of structures, e.g. buildings, bridges etc., $^{1}$H-NMR measurements based on constant magnetic fields, as known from the medical domain, are impossible on site. In contrast to the medical domain, single sided NMR works with a magnetic field gradient applied from the surface. So there are no size limits for investigated structures. In following, NMR results are shown as amplitude-depth diagrams. The amplitude is directly associated with the amount of hydrogen atoms in the measured volume. After a calibration process, the dimensionless amplitude can be used e. g. to determine the absolute moisture content (Keil et al. 2011). In order to reduce the measuring time, the investigated area was limited to a depth of 10 mm (ultimate measuring depth: 25 mm). The specimens were saturated by storing under water before each measurement to allow the determination of the hydrophobic layer thickness. Thus, the water was able to get behind the hydrophobic layer by penetrating the untreated sides of the specimen. Hydrophobic and non-hydrophobic areas can be differentiated in terms of the observed amplitude.

$^{1}$H-NMR focuses on the nuclei of hydrogen atoms. Thus, beside water, the hydrogen atoms of the hydrophobing agents within the concrete can also be visualised. By comparing measurements before and after application of the hydrophobing agents (Fig. 1 left) it is possible to follow the ingress over time, i.e. how deep the hydrophobing agent penetrates into the concrete (Fig. 1 left, 6,500–7,000 μm), and how the hydrophobing agent distributes. Seven days after application, the hydrophobised specimens were submerged in water to achieve maximum saturation and measured again. The results show that the penetration depth of the hydrophobing agent correlates with the inner side of the hydrophobic layer (Fig. 1 right). Control experiments showed that the thickness of the hydrophobic layer determined by breaking and spraying the specimen fits to an amplitude

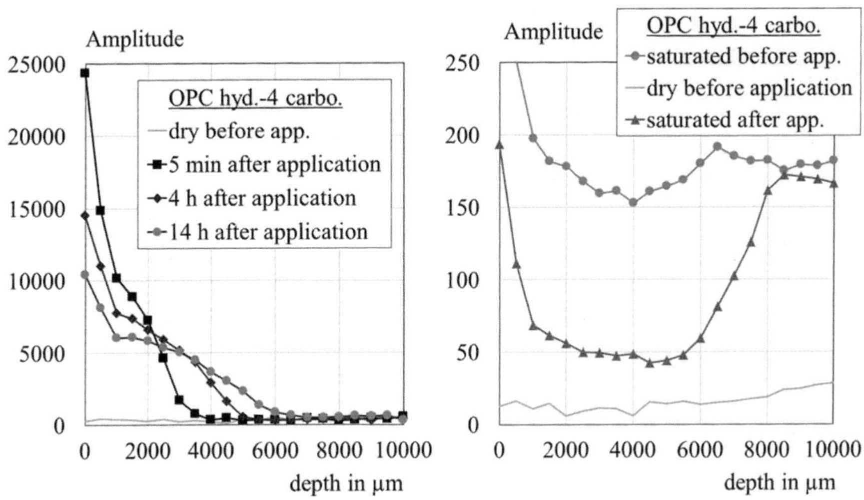


**Fig. 1** *Left*: amplitude-depth profile indicating the distribution of the hydrophobing agent hyd.- 4 applied on a carbonated OPC concrete; *right*: amplitude-depth profile indicating the water distribution before and after application of hydrophobing agent

of 100 (by using NMR settings to assess moisture, Fig. 1 right: ~7,000 μm). Due to the higher porosity in the first 2,000–3,000 μm below the mould-plain concrete surface the amplitude can be higher than 100 even under hydrophobic conditions. Hence it is possible to measure the thickness of the hydrophobic layer after application non-destructively by using single sided NMR.

## 3 Effects of Different Environmental Conditions on the Hydrophobic Layer Thickness

To identify the impact of relevant environmental factors on the hydrophobic layer thickness, series of specimens have been separately exposed to UV radiation, alkaline storage, and $CO_2$.

### 3.1 UV Radiation

Before the application of the hydrophobing agents, microstructural changes due to the hydration process had mostly finished and carbonation had reached a depth of more than 10 mm. The specimen shown in Fig. 2 left had been exposed to UV radiation for 27 days. In contrast to an unaffected hydrophobic surface the contact angle between droplet and concrete surface had changed from $\Theta < 90°$ to $\Theta > 90°$ i.e. the droplet spreads on the surface. Furthermore, the concrete surface below the water droplet had turned dark. Both effects indicate that the roll-off behaviour and

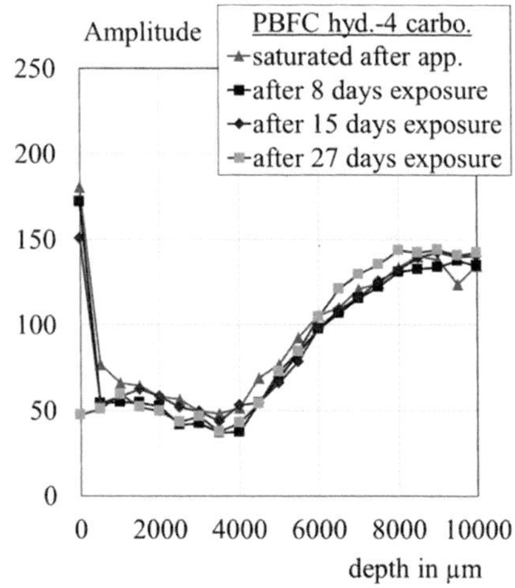


**Fig. 2** *Left*: water droplet on a concrete surface after 27 days of UV exposure; *right*: amplitude–depth profile indicating the water distribution after different times of UV exposure

the wettability have changed. However, during observation the droplet remained on the concrete surface. The water distribution shown in Fig. 2 right corroborates this. Furthermore, the profiles show that the UV radiation over time has no effect on the hydrophobic layer thickness. Thus, the impact of UV radiation is limited to the topmost surface layers. It should be mentioned that radiation which hits the concrete surface is transformed into thermal energy. The surface temperature of irradiated structures can reach approx. 80 °C. However, control experiments showed that temperatures below 250 °C have no effect on the hydrophobic layer thickness (Antons et al. 2012), which explains the unchanged hydrophobic properties of the hydrophobic layers below the surface.

## 3.2 Storing in Alkaline Solution

All hydrophobing agents listed in Table 1 are approved for the application on concrete by the guidelines for protection and repair of concrete parts of the German Board for Reinforced Concrete (DAfStb 2001). Their stability against alkaline attack has already been checked in the framework of the German approval procedure. In contrast to the standard test, the specimens were stored in alkaline solution for 3 months at 23 °C and for another 6 months at 80 °C. Additionally, specimens treated with methyltrimethoxy silane (hyd.-10), which is known to easily undergo hydrolysis in alkaline solution and is not approved for application on concrete, have also been investigated. Probably due to its small size, methyltrimethoxy silane is able to enter smaller pores, which are inaccessible by molecules with longer alkyl chains such as butyl or octyl. This is evidenced by the fact that despite water saturation of the specimen, the range between 2,500 and 5,500 μm shows a signal intensity equal to that of the dry specimen i.e. this area is essentially free of water. Such a drastic behaviour is not observed for any other hydrophobing agent (e.g. iso-octyltriethoxy silane treated specimen in Fig. 3 right). After exposure to alkaline solution, methyltrimethoxy silane treated specimens' show that water is able to penetrate into the previously hydrophobic area. The alkalis were able to partially destroy the hydrophobic layer. None of the approved hydrophobing agents have shown such a behaviour. Rather, the amplitude-depth curve after exposure to alkaline media matches that of before, except for a slightly decreased amplitude close to the surface (Fig. 3 right). This, however, is due to a continued reaction of the alkoxy silan with the pore surface.

## 3.3 Artificial Carbonation

There are significant differences between the microstructural changes caused by carbonation of OPC and PBFC concrete. In both concretes, calcium hydroxide (CH) reacts with carbon dioxide to calcium carbonate. In addition, the

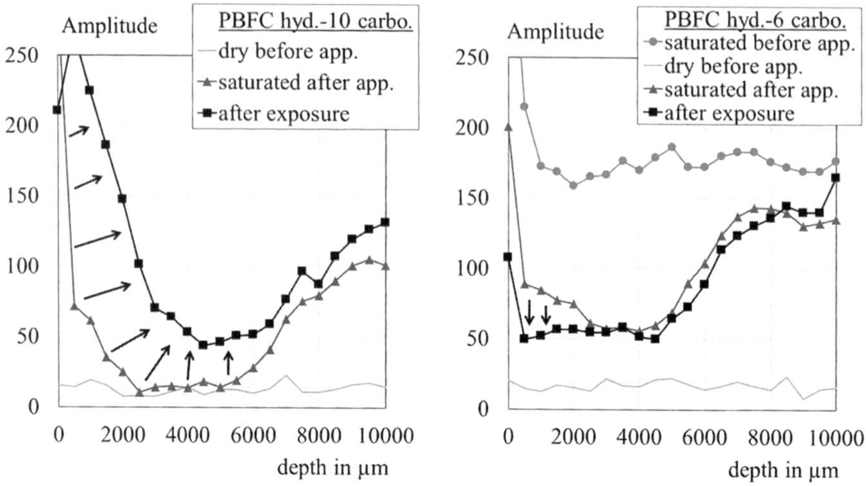


**Fig. 3** *Left*: moisture-depth profile of a specimen treated with methyltrimethoxy silane, dry before and saturated after application and exposure to alkaline solution; *right*: moisture-depth profile saturated or dry before application as well as saturated after application and exposure

transformation of calcium silicate hydrates (CSH) to calcium carbonate is also possible. However, the carbonation of CH is dominant for OPC concrete due to its rather large amount of free CH. In PBFC, the amount of CH available for carbonation depends on the quantity of blast furnace slag used in the cement mixture. At a content of more than 50 wt.-% blast furnace slag, the carbonation of CSH becomes significant (Bier 1988). In addition to calcium carbonate a second solid reaction product, namely silica gel, is built during the carbonation process of CSH. Upon carbonation, the pore size distribution of OPC based concrete shifts to smaller diameters. In contrast, during investigation the PBFC concrete reveals a shift to larger pore diameters with an increased proportion between 0.1 up to 2 µm after carbonation. This range of pore diameters is characteristic for the highly porous structure of the silica gel formed during the carbonation process of the CSH phases in the PBFC concrete (Bier 1988; Copuroglu et al. 2006). In addition, this newly generated, highly porous structure is hydrophilic.

The results show that carbonation of OPC concrete influenced neither the hydrophobic layer thickness nor its hydrophobic properties. In contrast, after carbonation of PBFC concrete the hydrophobic layer thickness decreased to a 1,000–2,000 µm wide area close to the surface (Fig. 4 left). Regarding the OPC results, the carbonation of CH and with it the crystallisation of calcium carbonate has not affected the hydrophobic layer. This indicates that the formation of silica gel during the carbonation of PBFC concrete might be responsible for the reduction of the hydrophobic layer thickness. In contrast to the UV exposure, carbonation has no influence on the roll-off behaviour (Fig. 4 right).

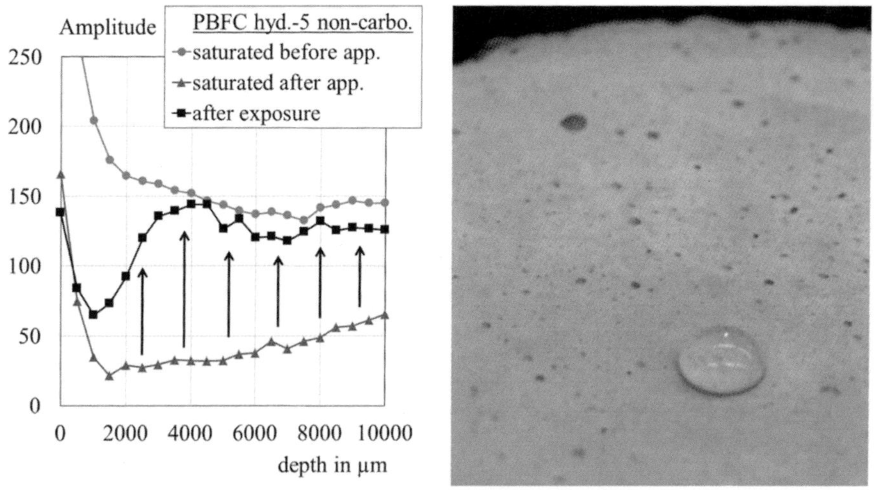


**Fig. 4** *Left*: moisture-depth profile saturated before and after application and saturated after exposure; *right* water droplet on a concrete surface after accelerated carbonation

## 4 Efficacy of the Hydrophobic Layer

Water ingress is the most important indicator for the efficacy of a hydrophobic layer. For the determination of the water ingress, the specimens' circumferences were sealed with epoxy resin and the specimens were stored in a footbath (Fig. 5 left). Thus, for water the only possible way to enter the specimen is through the hydrophobic layer. There are two different possibilities for water to cross the hydrophobic layer. Unaffected by the hydrophobic pore surfaces water vapour diffuses through the capillary system. If the vapour reaches non-hydrophobic surfaces in or behind the hydrophobic layer water can be adsorbed. The maximum amount of water which accumulates in these areas depends on the available pore surface area and the conditions affecting the absorption desorption equilibrium. In contrast to the vapour transport, the amount of water transported by capillary suction is much higher even if it is limited to a few spots within the hydrophobic area. However, the direct transport of liquid water is only possible if there are interconnected, non-hydrophobic surfaces bridging the hydrophobic layer.

In order to investigate the efficacy of a hydrophobic layer it is necessary to measure the rate and extent of water accumulation behind the hydrophobic layer, in an area unaffected by the hydrophobic treatment. To evaluate the accumulation over time, the amplitudes behind the hydrophobic layer were correlated to the water saturation state recorded of each specimen independently before the footbath storage (Fig. 5 left).

The specimens displayed in Fig. 5 right were hydrophobised in the un-carbonated (mode A) or carbonated state (mode B). All specimens were exposed to $CO_2$ after the hydrophobing agents were applied. As described in section

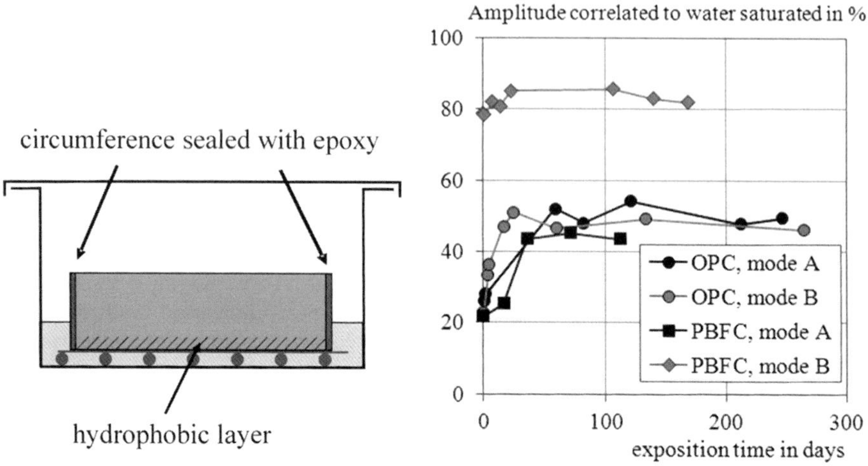


**Fig. 5** *Left*: specimen in footbath; *right*: amplitude behind the hydrophobic layer after footbath storage for OPC and PBFC specimens' mode A and mode B (modes A and B see text above)

"Artificial carbonation" only the previously un-carbonated PBFC specimens (mode B) showed changes in the hydrophobic layer.

As shown in Fig. 5 right, the water penetrates fast into the PBFC specimens' mode B. The amplitude measured after one day of exposure reaches nearly 80 % of the water saturated amplitude. The water ingress into OPC specimens' mode A and B as well as PBFC specimens' mode A is significantly slower and their amplitudes approach only approx. 40 % of the water saturated amplitude. After long exposure times the water ingress through the hydrophobic layer and the evaporation from the unsealed backside of the specimens are balanced. As mentioned above, only the hydrophobic layers of the PBFC specimens (mode B) were affected. The results suggest that water ingress through the unaffected hydrophobic layers is limited to the transport by vapour. The most probable explanation for the increased amplitude behind the affected hydrophobic layer of the carbonated PBFC specimens' mode A is that water is additionally transported by capillary suction.

## 5 Conclusions and Outlook

Questions concerning the durability of hydrophobic treatments can now be answered at a much more sophisticated level. The results show that, for the investigated hydrophobing agents approved for the application on concrete

- Alkalinity and elevated temperatures have no influence on the hydrophobic layer
- The effect of UV radiation is limited to the topmost surface layers of the exposed concrete surface.

Significant structural changes within the concrete have to occur in order to have an impact on the hydrophobic layer. In case of carbonation, the transformation of calcium hydroxide is without noticeable effect whereas the transformation of calcium silicate hydrates seems to have a major impact on the hydrophobic layer thickness and efficacy. With respect to the different types of cement used in this project it can be concluded that,

- The carbonation process of OPC has no effect on the thickness and efficacy of the hydrophobic layer
- The carbonation process of PBFC is affecting the hydrophobic layer thickness and efficacy.

Further investigations focus on the influence of the hydration process after application at early age. In addition the efficacy of specimens with different hydrophobic layer thicknesses will be evaluated.

**Acknowledgment** The authors thank the German Science Foundation (DFG) for funding the project "durability of hydrophobic treatments on concrete surfaces".

## References

Antons U, Orlowsky J, Raupach M (2012) A non-destructive test method for the performance of hydrophobic treatments. CRP Press/Taylor & Francis Group, London [u.a]. In: Proceedings of the 3rd international conference on concrete repair, rehabilitation and retrofitting (ICCRRR), Cape Town, 3–5 Sept 2012, ISBN: 978-0-415-89952-9A.T. 1990

Bier TA (1988) Karbonatisierung und Realkalisierung von Zementstein und Beton. Institut für Massivbau und Baustofftechnologie, Karlsruhe. In: Massivbau Baustofftechnologie Karlsruhe (1988), Nr. 4, Technische Universität, Karlsruhe, dissertation

Blümich B (2004) Essential NMR. Springer, Berlin. ISBN 3-540-23605-8

Copuroglu O, Fraaij ALA, Bijen JMJM (2006) Effect of sodium monofluorophosphate treatment on microstructure and frost salt scaling durability of slag cement paste. Cem Concr Res 36:1475–1482

DAfStb, Deutscher Ausschuß für Stahlbeton (2001) DAfStb-Instandsetzungs-Richtlinie: Schutz und Instandsetzung von Betonbauteilen. Teil 1: Allgemeine Regelungen und Planungsgrundsätze. Teil 2: Bauprodukte und Anwendung. Teil 3: Anforderungen an die Betriebe und Überwachung der Ausführung. Teil 4: Prüfverfahren

Keil A, Orlowsky J, Raupach M (2011) Einsatz eines mobilen NMR-Sensors als zerstörungsfreies Messsystem in der Bauwerkserhaltung: application of a mobile NMR sensor as a non destructive measuring tool to conserve buildings. Bautechnik 88(11):741–748. ISSN:0932-8351J

Meier SJ, Wittmann FH (2011) Recommendations for water repellent surface impregnation of concrete. Restor Build Monum 17(6):347–357

Polder RB, Borsje H, de Vries H (2001) Prevention of reinforcement corrosion by hydrophobic treatment of concrete. HERON 46(4):227–238

# Steel Corrosion Rate Measurements in FA Concrete Using Three Electrochemical Techniques

**M.P. López, J.M. Ortega, I. Sánchez, and M.A. Climent**

**Abstract** Blended cements are today commonly used for concrete structures built near the sea. When reinforced concrete is exposed to the atmospheric marine environment, it is necessary to consider steel corrosion caused by both, chloride ions and carbon dioxide. In this research, reinforced OPC and FA cement concrete samples have been tested in order to evaluate their comparative performance when exposed to chlorides, to $CO_2$ and to both agents acting together. The techniques used to evaluate the steel corrosion rates have been the polarization resistance, intersection method of polarization curves, and electrochemical impedance spectroscopy. The results indicate a similar behavior of FA and OPC concrete for samples exposed to $CO_2$. For the pure chloride exposure and the combined exposure regimes, the behavior depends on the type of binder. The polarization resistance technique appears to show some limitations when there is not enough electrolyte inside the pores.

**Keywords** Fly ash concrete • Corrosion • Electrochemical techniques

## 1 Introduction

The last edition of the Spanish Technical Code for structural concrete (Ministerio de Fomento 2010), introduces a new concept, the "Durability Limit State". According to this limit state, the time needed for an aggressive agent to produce a significant damage must be longer than the desired service life time multiplied by a safety factor. Besides, the same Spanish Standard states that under the same resistance conditions, sustainable materials must be used. This sustainability concept can be considered by calculating the Structure Sustainability Contribution Index. Thus, according to that standard, when designing a reinforced concrete standard, both, durability and sustainability must be taken into account.

It is well known that one of the main durability problems of reinforced concrete structures is steel bar corrosion. Two of the main aggressive agents that can cause

M.P. López (✉) • J.M. Ortega • I. Sánchez • M.A. Climent
Department of Civil Engineering, University of Alicante, Alicante, Spain
e-mail: mp.lopez@ua.es; jm.ortega@ua.es; isidro.sanchez@ua.es; ma.climent@ua.es



C. Andrade et al. (eds.), *Durability of Reinforced Concrete from Composition to Protection*, DOI 10.1007/978-3-319-09921-7_5

corrosion are chloride ion and carbon dioxide. On the other hand, previous works (Habert et al. 2010) show that over 80 % of $CO_2$ emissions of building material industry come from cement production. Regarding to that, blended cements are considered sustainable materials as they reduce the amount of clinker and so the $CO_2$ emissions.

In this work, reinforced concrete samples were prepared using a pozzolanic cement, CEM IV/B (V) 32.5 N according to European standard UNE-EN 197-1:2000. These specimens were subjected to chloride ion, $CO_2$ and both agents acting alternately, while testing the corrosion behavior of the embedded steel bars. Ordinary Portland Cement (OPC) reinforced samples have been used as a reference. With the objective to evaluate the corrosion state of the concrete bars, two different corrosion rate measurement techniques have been used, polarization resistance method and the intersection method of polarization curves, also known as Tafel slopes extrapolation technique. Results of both techniques have been compared and discussed. In addition, Electrochemical Impedance Spectroscopy (EIS) measurements have been carried out too.

## 2 Materials

Cylindrical concrete and mortar samples of 100 mm diameter and approximately 150 mm height were prepared. Dosage is shown in Table 1. For concrete and mortar samples made with fly ash, UNE-EN CEM IV/B (V) 32.5 N was used, which had a degree of replacement of clinker by FA between 36 % and 55 % in mass. As a reference material, ordinary Portland cement (UNE-EN CEM I 42.5 R) with 95–100 % of clinker was used.

As reinforcement, a ø 8 mm B500SD steel bar was placed vertically in the middle of the sample. After one week of curing inside a humidity chamber (95 % RH, 20 °C), cylindrical reinforced cores were drilled in order to get smaller samples of 50 mm in diameter and 20 mm of cover. Once the cores were cut, the experiment started.

**Table 1** Dosage of concrete and mortar samples

| Concrete samples | | Mortar samples | |
|---|---|---|---|
| Material | Mass (kg/$m^3$ concrete) | Material | Mass (g) |
| Cement | 350 | Cement | 450 |
| Aggregate [6–12 mm] | 714 | Sand | 1,350 |
| Aggregate [4–6 mm] | 489.5 | Water | 225 |
| Sand [0–4 mm] | 662.75 | w/c | 0.5 |
| Water | 175 | Sand/cement | 3:1 |
| Plasticizer | 1.5 % cement weight | | |
| w/c | 0.5 | | |

# 3 Experimental Set Up

## *3.1 Aggressive Exposure Environments*

Concrete samples were exposed to three different laboratory environments.

### 3.1.1 Chloride Ion Exposure Environment

Samples were exposed to a partial immersion (half of the height of the specimen) in a 3 % in mass NaCl solution. In this case, specimens were only removed from the solution during the corrosion rate measurements.

### 3.1.2 Carbon Dioxide Exposure

Periodic exposure (two times per week) to an atmosphere with a high $CO_2$ concentration. To this end the specimens were located in a closed desiccator. A pressurized tank of $CO_2$ was connected to the desiccator. A gas flow was provided from the pressurized tank to the non-pressurized desiccator. Once the desiccator was filled with $CO_2$, the flow was cut and samples were kept inside the desiccator until the $CO_2$ was slowly leaked. The $CO_2$ concentration was not measured in the carbonation chamber, while the temperature was that of the laboratory, i.e. $23\,°C \pm 2\,°C$, and relative humidity was approximately 65 %.

### 3.1.3 Chloride Ion and Carbon Dioxide Exposure

Alternate exposure to the chloride solution (partial immersion in 3 % NaCl), and to the $CO_2$ cycles. The frequency was the following: one week in chloride immersion and one week in laboratory exposure. During the laboratory exposure week, samples were subjected to two carbon dioxide cycles.

## *3.2 Procedures*

Corrosion rate measurements were performed with the objective to study the corrosion state of the bars. In order to do that, two corrosion rate measurement techniques were used, linear polarization resistance method (LPR) and intersection or Tafel extrapolation method (ITE). A comparison of results from both techniques was carried out. In addition, electrochemical impedance spectroscopy (EIS) measurements were carried out.

### 3.2.1 Corrosion Rate: Tafel Extrapolation and Polarization Resistance Methods

For the ITE, the device used was an AUTOLAB potentiostat/galvanostat and the software "General Purpose Electrochemical". Samples are subjected to a potential scanning around its $E_{corr}$ value. The graphical representation of the Intensity-Voltage values allows us to draw the anodic and cathodic slopes (Fig. 1). The intersection of those lines provides the $I_{corr}$ value. The scanning rate used was 0.001 V/s. The mean value of the potential range applied was 0.150v. The ohmic drop compensation was removed automatically by the device. Figure 1 shows the intersection of the Tafel slopes for a concrete sample of OPC after 80 days exposure to chloride attack.

For the LPR, a scanning potentiostat 362 from EG&G Instruments was used. In this case, as it was necessary to carry out iR compensation, an oscilloscope HAMEG HM 203-7 (20 MHz) was used. The scanning rate was 0.5 mV/s and the potential scanning range [$E_{corr}$ – 10 mV, $E_{corr}$ + 10 mV]. A value of B = 26 mV was taken (Andrade and González 1978) for the calculation of the $i_{corr}$ values.

For both techniques the counter electrode was a stainless steel mesh, surrounding the concrete specimens. The steel bar was connected to the working electrode and finally, a Ag/AgCl electrode was used as a reference electrode. This arrangement can be observed in Fig. 2. As the perturbation applied in the LPR method is low, and

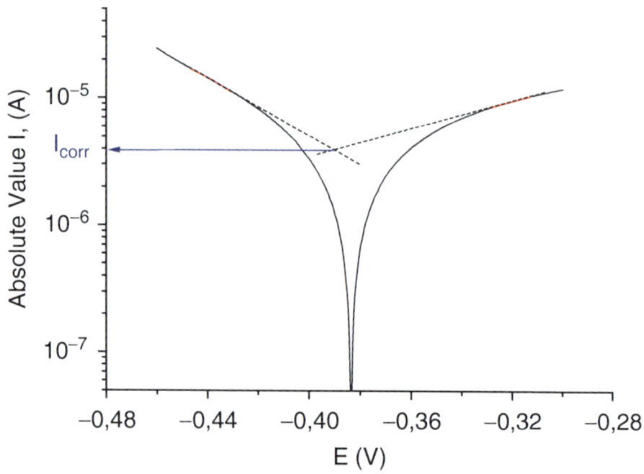


**Fig. 1** Intersection of Tafel slopes extrapolation for a CEM I concrete sample after 80 days exposure to chloride ions

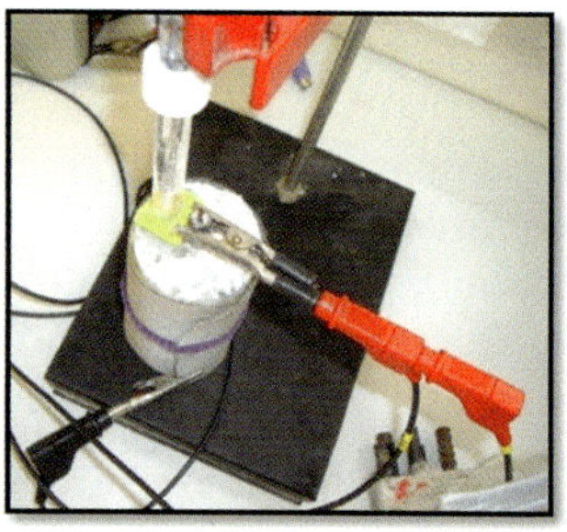

**Fig. 2** Experimental set up

so it can be affected by other electrical devices, the measurements were performed inside a Faraday cage.

### 3.2.2 Electrochemical Impedance Spectroscopy

For the EIS measurements, an AUTOLAB, potentiostat/galvanostat with Frequency Response Analysis module was used (AC). The frequency range selected was 1 mHz – 1 MHz (0.001–$10^6$ Hz). The setup is the same as that used for LPR and ITE (Fig. 2). The equivalent circuit employed for the electrochemical impedance spectra fitting is shown in Fig. 3.

In this circuit, $R_s$ is the electrolyte resistance, related to the porosity of the material (Andrade et al. 2001a, b). Capacitance $C_0$ and resistance $R_0$ are associated to redox processes in the oxide layer (Morozov et al. 2013). $R_p$ is the charge transfer resistance and $C_{dc}$ represents the double layer capacitance. The measurement fitting was done by using Zfit software (de Vera Almenar 2007). The Differential Impedance Analysis (Stoynov 1989) carried out (Fig. 4), showed the presence of two time

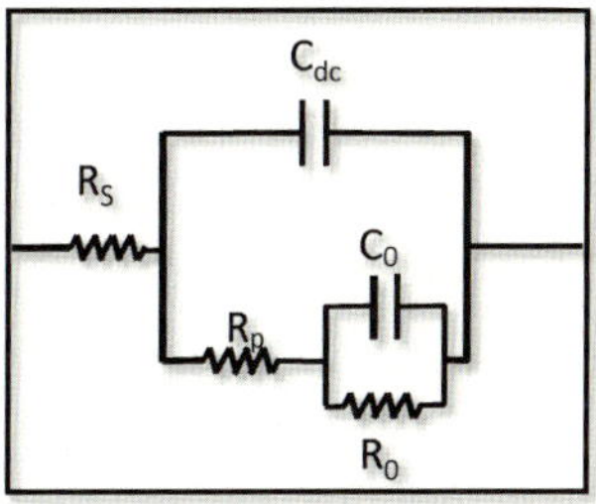


**Fig. 3** Equivalent circuit

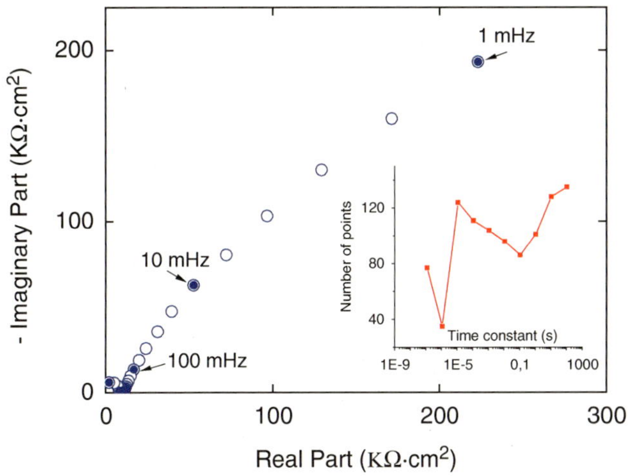


**Fig. 4** Impedance Spectrum of a FA concrete sample subjected to chloride contamination after 30 days exposure and differential impedance analysis

constants within the measured frequency range. The value of those time constants can be modified as the spectrum does. For a fly ash concrete specimen, after 30 days exposure to NaCl partial solution, the time constants values are $T = 2 \cdot 10 - 5$ s and $T = 3{,}000$ s which is equivalent to 50 kHz and 0.3 mHz.

# 4 Results and Discussion

## *4.1 Intersection Method Versus Polarization Method*

A comparison of the results of corrosion rate values and corrosion potential obtained from LPR and ITE was carried out. It can be observed how $E_{corr}$ and $I_{corr}$ values are different depending on the method used (Figs. 5, 6, 7, and 8). Those

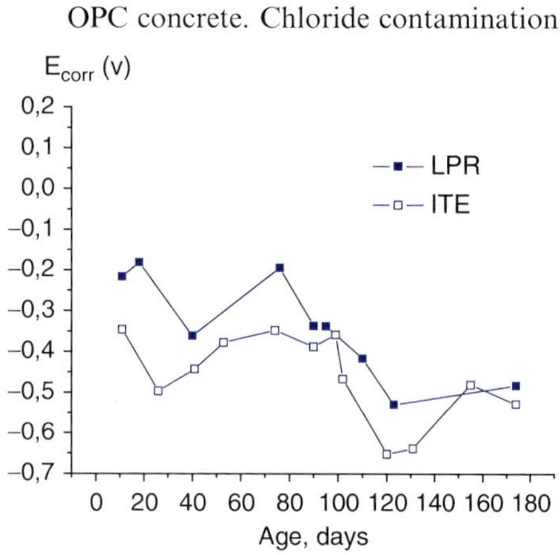


**Fig. 5** Corrosion potential values obtained for a representative OPC concrete sample exposed to chloride contamination

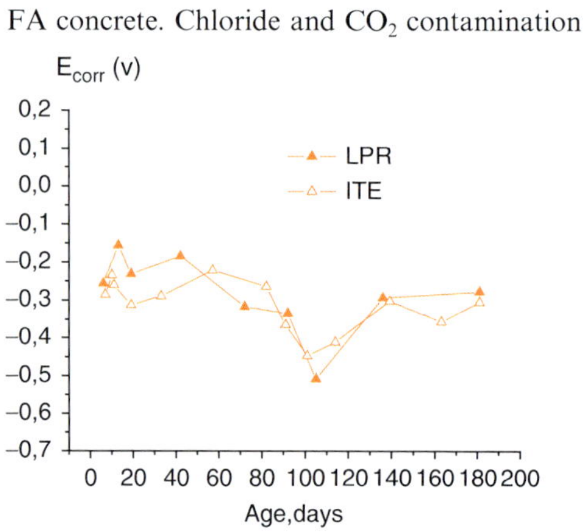


**Fig. 6** Corrosion potential values obtained for a representative FA concrete sample exposed to alternately chloride and $CO_2$

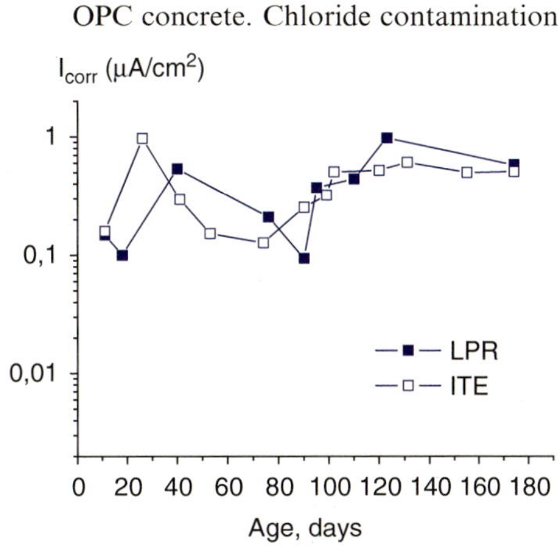


**Fig. 7** Corrosion rate values obtained for a representative OPC concrete sample exposed to chloride contamination

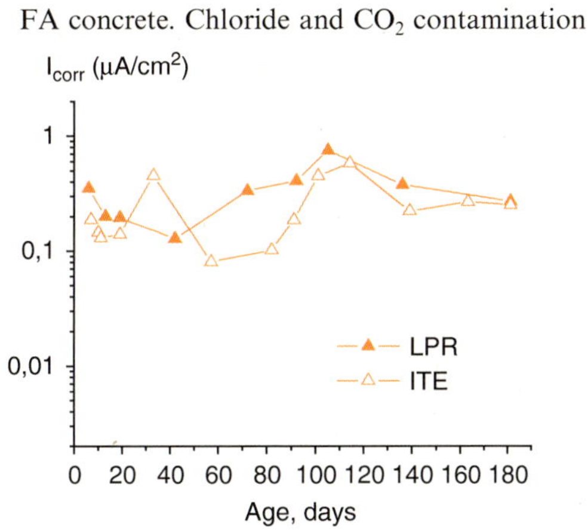


**Fig 8** Corrosion rate values obtained for a representative FA concrete sample exposed to alternately chloride and $CO_2$

differences do not follow a specific trend. Despite differences between the corrosion rate values, they are within the same order of magnitude.

Besides, when samples were too dry, that is, for the samples subjected to $CO_2$ contamination, the LPR measurements appeared to be unstable. It was difficult to obtain a stable value of $E_{corr}$, needed previously to the potential scanning. In those cases it was almost impossible to obtain a reliable value of $I_{corr}$. When concrete is very dry the corrosion value is expected to be negligible. Due to the difficulty to obtain $I_{corr}$ values in that cases, for the comparison of the behavior of FA concrete and OPC concretes only results from ITE were used.

Figures 9, 10, and 11 show the corrosion rate values (ITE method) of OPC and FA concrete samples subjected to the three different exposure environments.

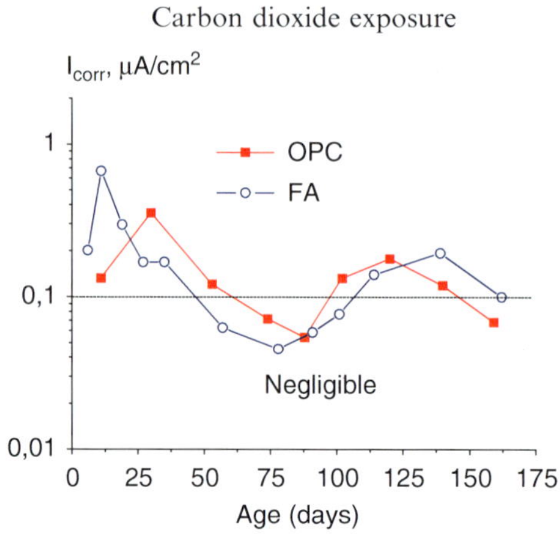


**Fig. 9** $I_{corr}$ values for samples subjected to $CO_2$ exposure

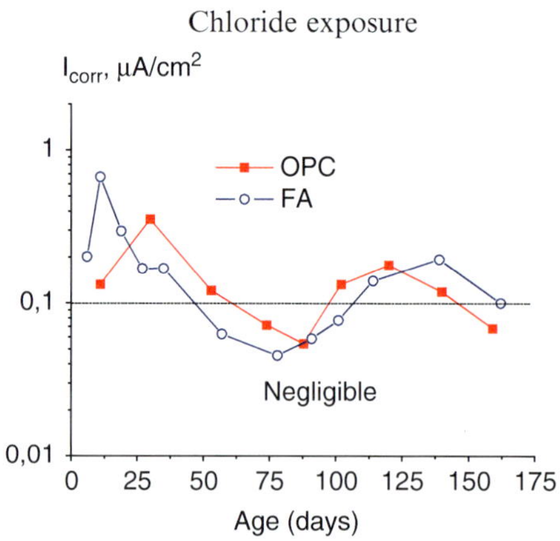


**Fig. 10** $I_{corr}$ values for samples subjected to chloride exposure

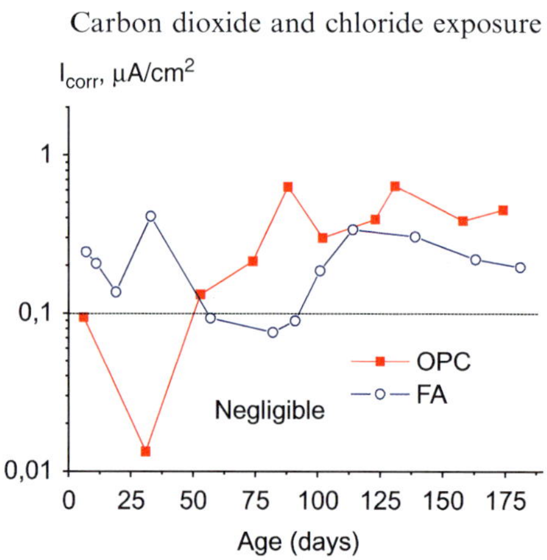


**Fig. 11** $I_{corr}$ values for samples subjected alternately to $CO_2$ and chlorides exposure

As it can be seen in Fig. 9, OPC concrete and FA concrete seem to behave similarly when they are exposed to carbon dioxide. As well, the $I_{corr}$ values obtained in this case are low, as explained before, that is below 0.2 μA/cm$^2$ (Zornoza 2007). When samples are exposed to chloride ions, $I_{corr}$ values at early ages are high for both types of concrete (Fig. 10). However, at the age of 75 days, FA concrete shows lower values of corrosion rate, which can mean whether this type of concrete provides better protection against corrosion, likely due to its more refined pore structure (Sánchez et al. 2011), or its corrosion is more localized. $I_{corr}$ values of OPC concrete between 0.5 and 0.6 μA/cm$^2$ show a moderate corrosion state of the steel bars. Finally, when samples are subjected to the action of both aggressive agents (Fig. 11), chlorides and carbon dioxide, under the conditions of the present study, chloride ions seem to be the dominant agent since the results are similar to those obtained when samples were exposed only to chlorides. There are not so high differences between OPC and FA concrete as it would be expected (Polder and Peelen 2002). This could be due to a short curing time. Pozzolanic reactions of the FA take longer than the clinker hydration reactions, and probably the protective refined pore structure characteristic of the FA concrete had not been developed completely when the test started.

Values of corrosion potential (Figs. 12, 13, and 14) show as well no significant differences between OPC and FA concrete (ITE method). However, there is an evidence of the different consequences of the three exposure environments. The carbon dioxide exposure gives $E_{corr}$ values over −0.2 V, which means that the probability of corrosion is very low (ASTM C 876-91). The lowest values of $E_{corr}$ are given for the $CO_2$ and chloride exposure, what reveals that this is the most aggressive of the three conditions studied in this work.

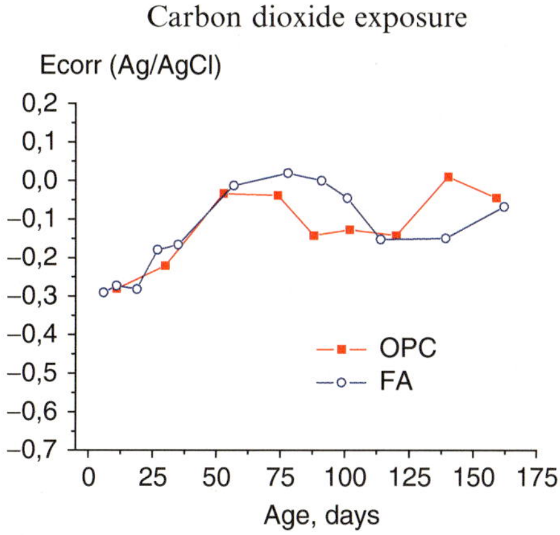


**Fig. 12** $E_{corr}$ values for samples subjected to $CO_2$ exposure

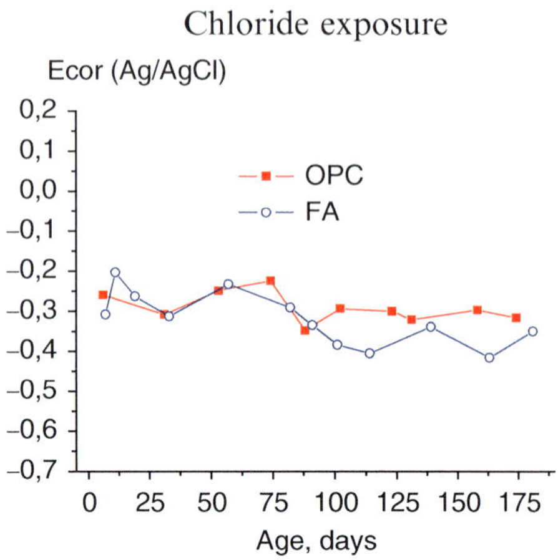


**Fig. 13** $E_{corr}$ values for samples subjected to chloride exposure

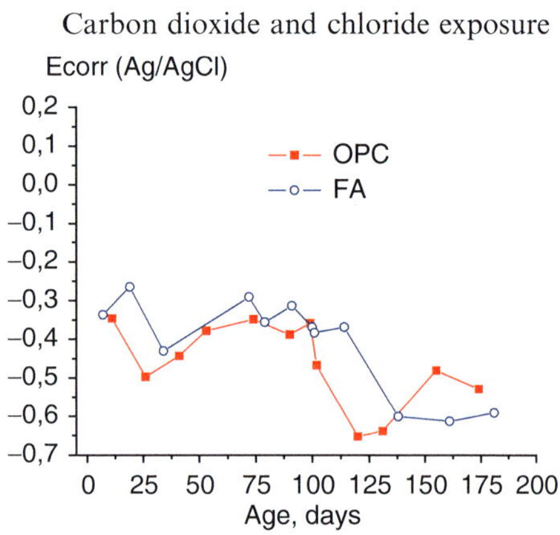


**Fig. 14** $E_{corr}$ values for samples subjected alternately to $CO_2$ and chlorides exposure

## *4.2 Electrochemical Impedance Spectroscopy (EIS)*

The first thing that was observed from EIS results, was the fact that $R_s$ was highly affected by induction phenomena and so no certain values of that parameter could be obtained. Besides, $C_0$ and $R_0$ parameters were difficult to fit especially when ion diffusion processes were considerable. Thus, only $R_p$ values are shown in this work. Comparing polarization resistance values obtained from ITE method and charge transfer resistance values obtained from EIS (Fig. 15) it is possible to see a relation between both parameters but only for samples exposed to $CO_2$.

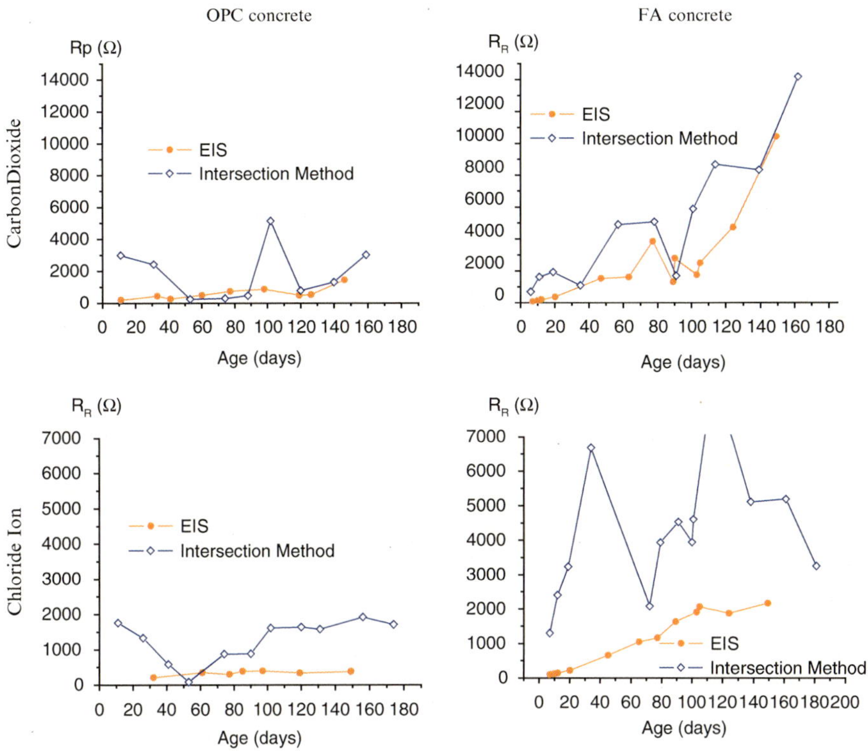


**Fig. 15** Polarization resistance and charge transfer resistance

# 5 Conclusions

OPC concrete behaves similarly to concrete made with FA addition when it is exposed to $CO_2$ contamination, with the particular conditions of this work. $I_{corr}$ values are insignificant due to the drystate of the samples. The absence of water inside the pores prevents the development of the oxidation/reduction phenomena at a significant rate.

After 75 days exposure to chloride ions, FA concrete shows a slightly lower corrosion rate values than OPC concrete. However, due to the short curing time given to the specimens, the expected beneficial effect of that addition hasn't been reflected. When samples were exposed to the combination of carbon dioxide and chlorides, the effect of chloride ions was dominant.

The value of $R_p$ obtained by EIS is similar to the $R_p$ value obtained by the ITE method for carbon dioxide exposure. The specific potential limit value that establishes the Rp(ITE)-Rp(EIS) relationship, should be studied. For corrosion due to chloride, LPR and ITE gave good correspondence.

**Acknowledgements** Authors would like to thank the Spanish "Ministerio de Economía y Competitividad" former "Ministerio de Ciencia e Innovación" and "FEDER" for the financial support through projects BIA 2010-20548 and BIA 2011-25721. M. P. López is indebted to the "Ministerio de Economía y Competitividad" for a fellowship of the "Formación Personal Investigador (FPI)" programme (reference BES-2011-046401).

## References

Andrade C, González JA (1978) Quantitative measurements of corrosion rate of reinforcing steels embedded in concrete using polarization resistance measurements. Mater Corros 29(8):515–519

Andrade C, Bolzoni F, Cabeza M, Nóvoa XR, Pérez MC (2001a) Measurement of steel corrosion in concrete by electrochemical techniques: influence of the redox processes in oxides scales. In: Bonora PL, Deflorian F (eds) Electrochemical approach to selected corrosion and corrosion control studies. Londres, IOM Communications Ltd, London

Andrade C, Keddam M, Nóvoa XR, Pérez MC, Rangel CM, Takenouti H (2001b) Electrochemical behavior of steel rebars in concrete: influence of environmental factors and cement chemistry. Electrochimica Acta 46:3905–3912

ASTM Standard Test Method for Half-Cell Potentials of Uncoated Reinforcing Steel in Concrete C 876–91 (Reapproved in 1999)

Ministerio de Fomento de España (2010) Instrucción de Hormigón Estructural EHE-0 (Spanish standard)

de Vera Almenar G (2007) Zfit manual. Department of Civil Engineering, University of Alicante. Alicante, Spain

Habert G, Billard C, Rossi P, Chen C, Roussel N (2010) Cement production technology improvement compared to factor 4 objectives. Cement Concr Res 40:820–836

Morozov Y, Castela AS, Dias APS, Montemor MF (2013) Chloride-induced corrosion behavior of reinforcing steel in spent fluid cracking catalyst modified mortars. Cem Concr Res 47(4):1–7

Polder RB, Peelen WHA (2002) Characterisation of chloride transport and reinforcement corrosion in concrete under cyclic wetting and drying by electrical resistivity. Cem Concr Compos 24:427–435

Sánchez I, López MP, Ortega JM, Climent MA (2011) Impedance spectroscopy. An efficient tool to determine the non-steady-state chloride diffusion coefficient in building materials. Mater Corros 62(2):139–145

Stoynov Z (1989) Structural spectral analysis of electrochemical impedance. Electrochimica Acta 8(34):1187–1989

Zornoza E (2007) El papel del catalizador usado de craqueo catalítico (FCC) como material puzolánico en el proceso de corrosión de armaduras de hormigón (The role of wasted catalytic cracking catalyst as pozzolanic material in the corrosion process of concrete steel reinforcement). Doctoral thesis, Polytechnic University of Valencia (in Spanish)

# Cold Drawn Steel Surface Analysis in Contact with Saline Solution: Analysis Using Electrochemical Atomic Force Microscopy

**Alicia Pachón, Javier Sánchez, Carmen Andrade, Esperanza Menéndez, and José Fullea**

**Abstract** High strength steels in concrete structures used as reinforcement are usually passivated due to the physical barrier that concrete provides and by a protective passive film developed on the surface by the high alkaline media. Chloride ions can break the passive film. These ions can be found dissolved in the water or in structures in marine environment.

This work is focused on the study, by Electrochemical Atomic Force Microscopy (EC-AFM), of the early stages of corrosion of such steels during exposure to a sodium chloride solution in order to better understand the evolution of the dissolution of steel phases. Additionally, it is estimated the corrosion rate by topographical images and simultaneously by means of measuring the corrosion rate through the linear polarization technique performed in an electrochemical cell specially developed in the AFM microscope. The results show the evolution of the topographical profiles which are compared during some period of time with the electrochemical results.

**Keywords** Reinforcement steel • Corrosion and atomic force microscopy (AFM)

## 1 Introduction

Concrete is an alkaline media which protects the steel from corrosion by two natural ways: it represents a protective physical barrier and the high alkalinity of the concrete pore solution develops a passive layer on the steel surface which may remain forever. However the penetration of the carbonation front or chlorides ions can disrupt this passive film developing active corrosion.

On the other hand, in the last years, nano and micro scale evolution surface studies are becoming important in order to get deeper knowledge about corrosion (Bertrand 2000; Cui 2006; Datta 2008; KamachiMudali and Katada 2001;

A. Pachón (✉) • J. Sánchez • C. Andrade • E. Menéndez • J. Fullea
Department of Durability, Eduardo Torroja Institute of Construction Science, Madrid, Spain
e-mail: apachon@ietcc.csic.es

 
C. Andrade et al. (eds.), *Durability of Reinforced Concrete from Composition to Protection*, DOI 10.1007/978-3-319-09921-7_6

Li and Meier 1998; Nagarajan and Rajendran 2009; Rongguang 2004; Sánchez 2008; Singh 2009; Wang and Kido 2006; Xu et al. 2002; Yanliang et al. 2008; Zhang 2005).

Atomic Force Microscopy (AFM) is a materials surface characterization technique that provides micro and nano structural information in high resolution and in a three-dimensional way. This technique enables to obtain topographic images "in situ" in liquids, under alkaline or neutral conditions and different concentrations of aggressive agents contained in the solution.

This microscope enables the simultaneous coupling of non-destructive electrochemical techniques, like Linear Polarization Resistance (LPR) which involves a polarising of +/−20 mV in every test.

It is combined electrochemistry techniques and Topography images, which are obtained simultaneously in saline solutions.

Present work has the goal to study the anodic dissolution of cold drawn steel wires with eutectoid composition and with the original structure of manufacturing. The experiments are carrying out in solutions which try to simulate the concrete and the sea water composition. The methodology combines the atomic force microscopy and linear polarization resistance techniques.

# 2 Experimental

## 2.1 Material

The working electrode was cold drawn steel Y1670C. The chemical composition of this material is given in Table 1.

The iron composition was also calculated by Florescence Diffraction Ray-X (FDRX). The semi quantitative value was 98.73 % of Fe in the elemental Chemical Analysis which was found consistent with the result of the chemical analysis.

Metallographic studies show that the cold drawn steel microstructure is lamellar perlite throughout the transversal section. A micro structural anisotropy caused by the drawing processing is observed. The pearlite colonies are oriented in the direction of drawing (Fig. 1).

## 2.2 Sample Preparation

A small cold drawn steel cylinder is embedded in epoxy resin plus some hardener. In order not to modify the microstructure, a cold sample is cut in thin sheets in

**Table 1** Results of the wet chemical analysis of the steel shavings

| C % | Mn % | Si % | P % | S % | Cr % |
|---|---|---|---|---|---|
| 0.82 | 0.65 | 0.17 | <0.015 | 0.022 | 0.28 |

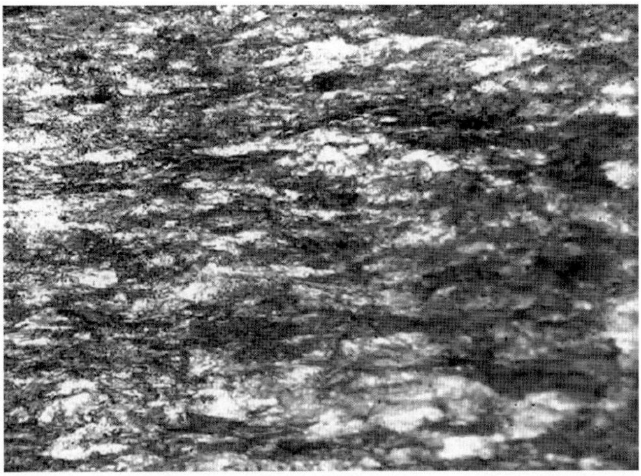

**Fig. 1** Micrography ×500 (500 increments) longitudinal section of cold drawn steel

adjusted dimensions for the AFM sample plate. Because the AFM technique requires extremely flat surfaces with lateral resolution around one micron, samples were polished using some SiC emery papers and secondly with micro crystals of diamonds of 3 and 1 micros. They are rinsed, after each diamond paste polishing cycle, with a soap-water mixture, then with ethanol and finally they are dried with a hairdryer.

## 2.3 Equipment

Electrochemical force microscopy studies were performed using an AFM 5500 from Agilent technology. The AFM surfaces are analysed in contact mode with a Hydra 6 V–100NG tip with the following characteristics; Low stress SiN y Tip: Si, L = 100 μm W = 18 μm; T = 0.8 μm. V shape. Tip ratio: <10 nm, H; 4–6 μm, f = 66 KHz, F = 0.029 N/m. These probes are designed for imaging soft samples in no contact mode or tapping mode but can also be used for imaging soft samples in contact mode. These probes have a short, narrow, and V-shaped cantilever. The gold coating on the reflex side allows for improved signal strength in fluid environments. The triangular shape of cantilever decrease friction forces when the tip touches surfaces along the sweep.

For the simultaneous LRP measurements, an electrochemical cell is inserted in the sample plate which has three electrical connections: an Ag/AgCl reference electrode with salt bridge, a steel sample working electrode and a platinum counter electrode. Electrochemical values are measured with an external ACM Potentiostat/Galvanostat (Fig. 2).

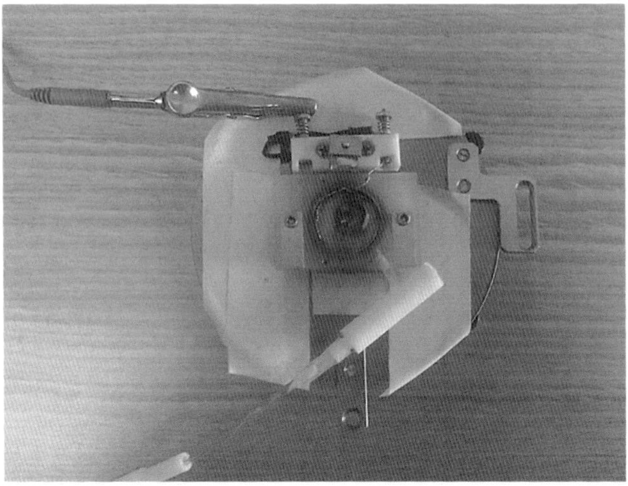

**Fig. 2** Electrochemical cell with three electrodes

Samples were studied in two different neutral salt solutions; 0.005 M and 0.01 M NaCl with a pH of 7.5 and in one alkaline saline solution; 0.01 M NaCl with a pH of 12.8.

## 2.4 Techniques Used

### 2.4.1 Atomic Force Microscopy

Atomic Force Microscopy uses a scanner made of piezoelectric material that moves a small cantilever with a very small tip fixed to it. There are three basic working modes by AFM. First one is "contact mode". It is when the tip is in permanent contact along the sweep surface. Second one is "tapping mode", when the tip tap the sweep surface. And the last one is "no-contact mode" when the tip does not touch the surface but it is much closed to it (see Fig. 3).

In present work the contact mode and constant force setting is used. The force applied has been the lowest possible in order to avoid the tip off the surface during scanning. In this way we try not booting surface material despite working in contact mode.

### 2.4.2 Electrochemical Technique

Linear Polarization Resistance (LPR) is used to measure the corrosion rate. This procedure is based on the fact that polarization curves next to the $E_{corr}$ in a small interval are practically straight lines. The slope of potential over current has an

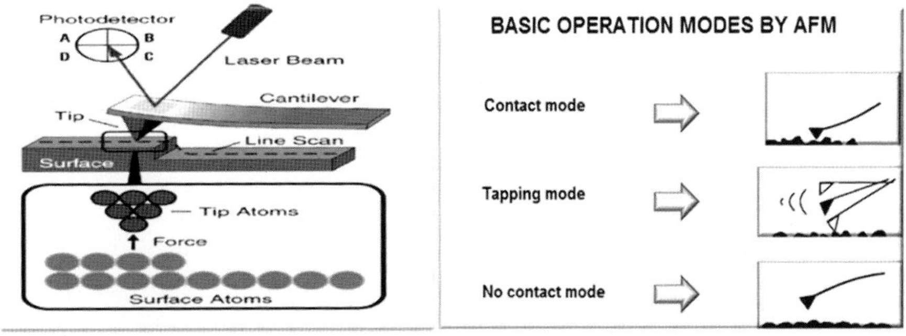


**Fig. 3** Diagram of atomic force microscopy and three basic working modes

inverse relation to related the corrosion rate. In practice a small polarization of +/−20 mV is applied in the surroundings of generally Open Circuit Potential (OCP) with a scan rate of 10 mV/min. The corrosion rate is calculated using the simplest formula of Stern and Geary:

$$Icorr = \frac{B}{Rp} \tag{1}$$

Where B = constant, Icorr = rate corrosion, Rp = linear polarization resistance.

The no-destructive nature of the test allows to repeat the measurement in the same reinforcement for evaluating rate corrosion along the time.

According to this methodology, it is carried out AFM and LRP tests simultaneously during 4 h. LPR measurement is performed every 15 min.

## 3 Results and Discussion

In Fig. 4 are given topographical images of 5 × 5 microns which show the evolution of cold drawn steel surface in a solution of 0.01 M NaCl and pH 7.5. At the beginning it can be appreciated the grooves resulting from the polishing. From the figure after 15 min of contact with the chloride solution it is already appreciated the progressive dissolution of ferrite phases which leads to leave standing out isolated cementite one.

The following images (Fig. 5) show topographical images of 5 × 5 microns with the evolution of cold drawn steel in a solution of 0.01 M NaCl and pH 12.8. In this case the ferrite phases also are only very superficially dissolved, and therefore it cannot be identified a clear attack.

Cold drawn steel has a lamellar pearlite composition of cementite and ferrite. Ferrite is mainly alpha-iron whereas cementite is mainly iron carbide $F_3C$ with a mass proportion of 6.66 % carbon and 93.3 % iron. In Fig. 6 it is reproduced the interpretation of what has been observed. At the beginning (Fig. 6a) at the surface

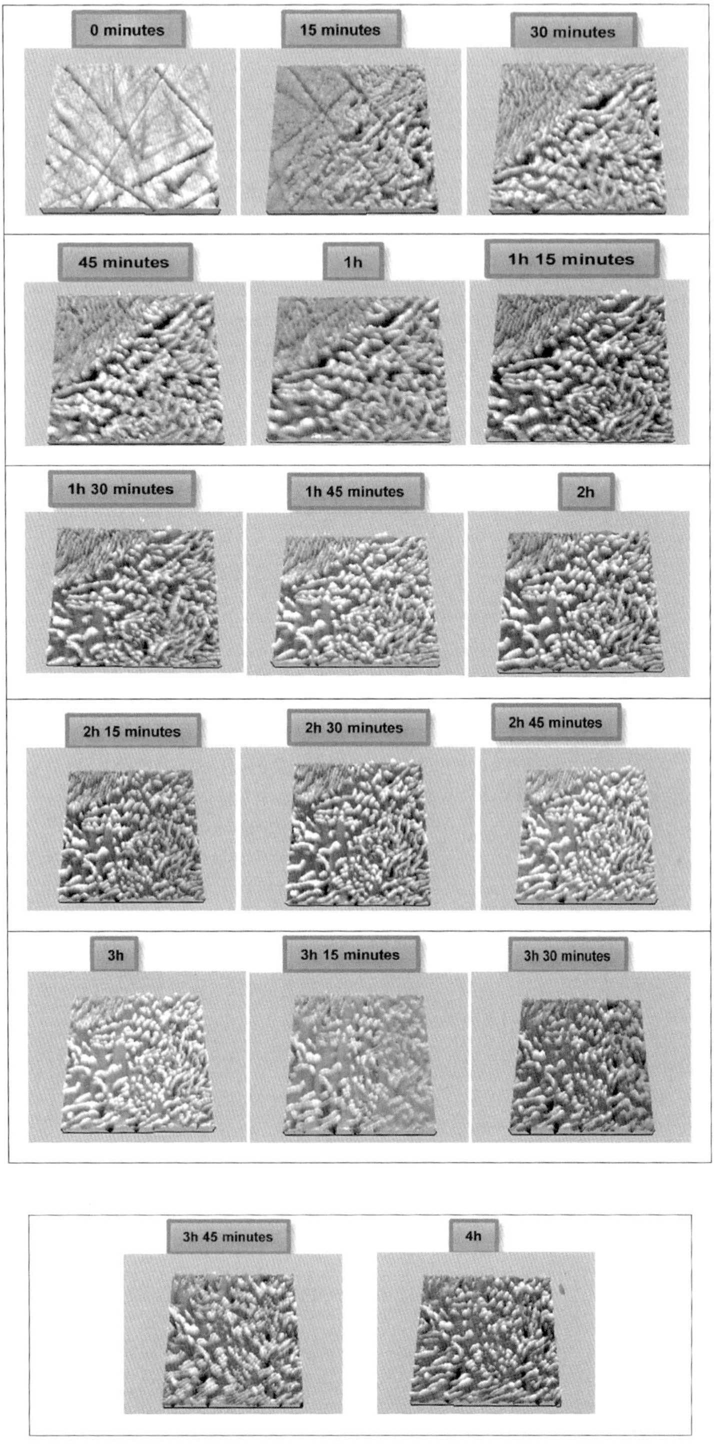


**Fig. 4** Cold drawn steel evolution into solution of 0.01 M in NaCl and pH 7.5

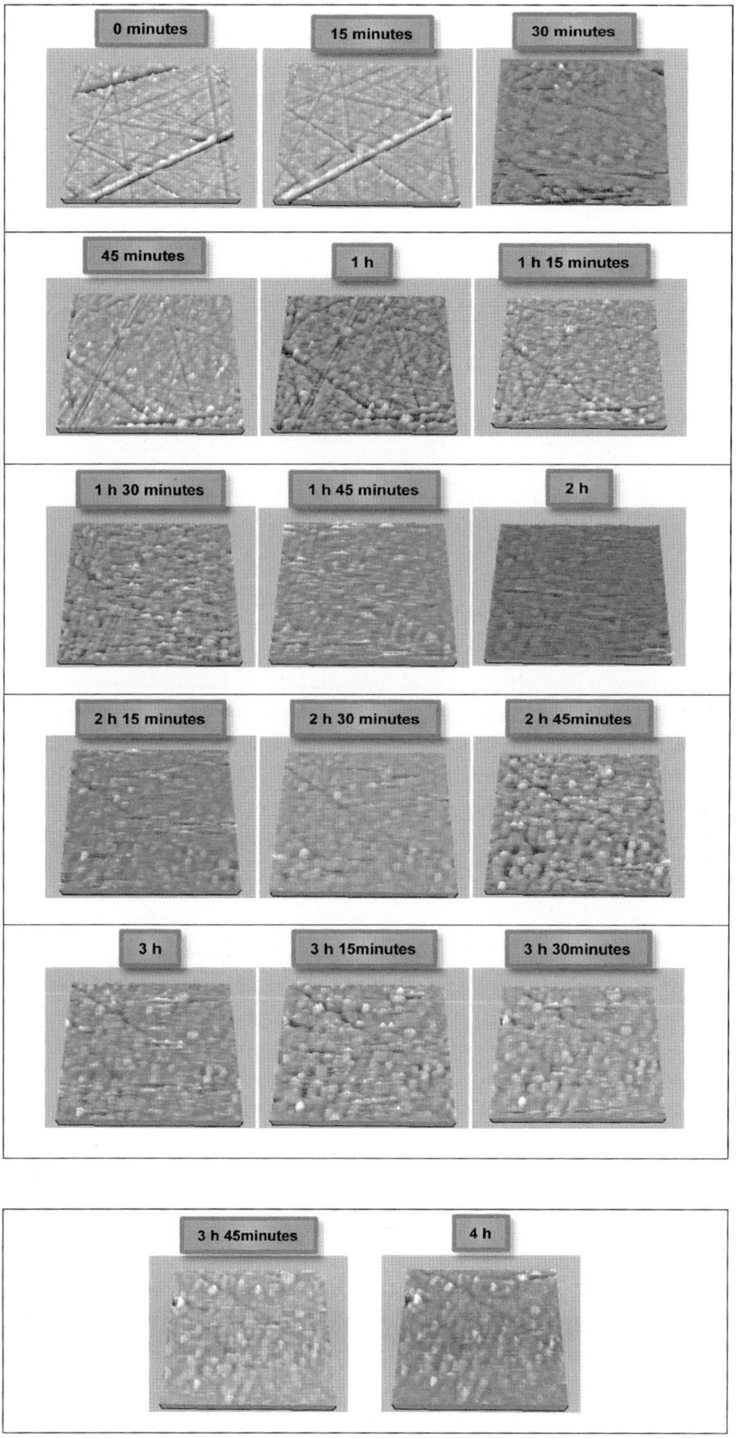


**Fig. 5** Cold drawn steel evolution into solution of 0.01 M in NaCl and pH 12.8

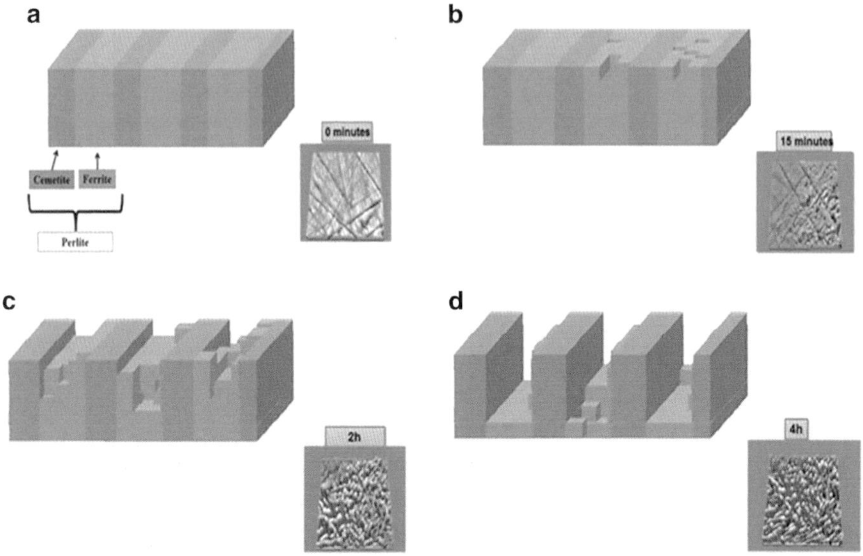


**Fig. 6** Model and experimental results diagram

emerge both phases but when the dissolution evolves, the ferrite is corroding and it began to disappear (Fig. 6b–d). Then, it can be assumed that ferrite is acting as anode and cementite phase as cathode (Fig. 6c). During the experiment it seems that cementite does not suffer from any corrosion (Fig. 6d).

In order to check this observation by other technique it has been tried to derive the amount of material lost from the topographical profiles (roughness study) observed.

## *3.1 Calculation of the Average Topography Depth*

From each topographical image by AFM a histogram is obtained by PicoImage Agilent software. These histograms provide the different depths from the whole area versus the percentage of each which are detect along the entire swept surface, as shown in Fig. 7.

Taking these values as reference, the average topography depth is calculated and plotted as a point in a graph of mean topographical depth (nm) versus time (s).

## *3.2 Calculus of Averaged Depth by Means of Liner Polarization and Faraday's Law*

Every 15 min a linear polarization resistance measurement is made and in the meantime, by AFM, topographic images of the steel surface are obtained inside the solution for a total time of 4 h.

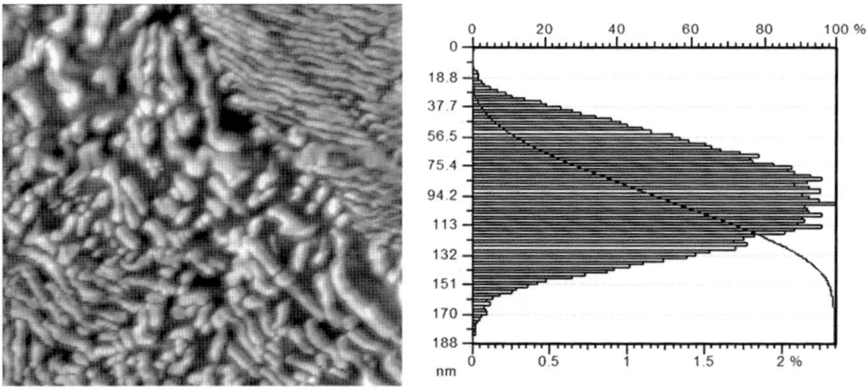


**Fig. 7** Example of a bar histogram which corresponds to a topography image after 1 h and 15 min subjected to a salt solution of 0.01 M NaCl and pH 7.6

Faraday's law states that the passing of 96493 coulombs (1 Faraday) through a metal/solution interface causes the dissolution or deposition of one equivalent substance. This law allows calculation of the corroded quantities of a metal (by weight) according to the corrosion current:

$$\frac{\Delta \mathrm{w}}{\frac{Pm}{\mathrm{z}}} = \frac{i * \Delta t}{F} \tag{2}$$

Where Δw = lost mass (g), Pm = molecular weight (g/mol), i = current (A), Δt = time of current flow across the cell (s), F = Faraday constant 96493 coulombs/equivalent, z = number of transferred electrons.

Then, the corrosion rate is calculated from each LPR measurement by Eq. 1. And using Faraday's law, Eq. 2, the mass loss is calculated.

Mass loss and iron density are used for calculating the lost volume, according to Eq. 3. Finally, with total area of the working electrode the mean depth of corrosion is obtained through Eq. 4. This averaged depth is called Faraday's depth.

$$\Delta w = \rho * \Delta V \tag{3}$$

Where ΔV = lost volume, ρ = iron density

$$\Delta P = \frac{\Delta V}{At} \tag{4}$$

Where ΔP = averaged depth or Faraday depth, At = total working electrode surface (sample).

In Fig. 8 it is plotted, for each testing time, the averaged depth from polarization resistance measurements and that from the images of topography by AFM. In this

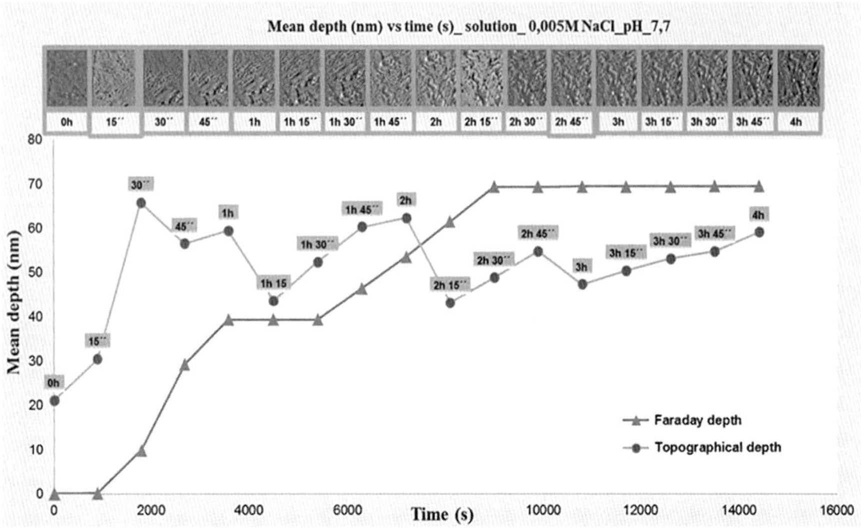


**Fig. 8** Topographical and Faraday depths compared

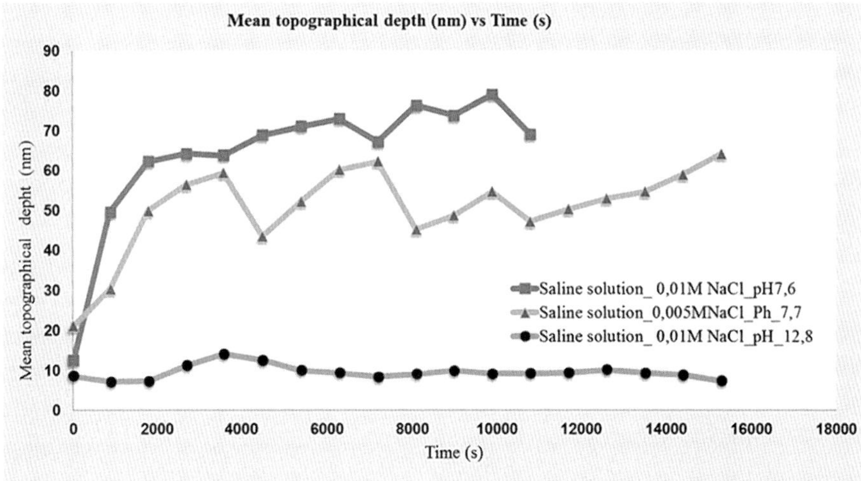


**Fig. 9** Topographical depths diagram for different saline solutions

case, the experiment was carried out in 0.005 M sodium chloride solution and pH 7.6. Both lines have the same tendency and their values are close enough despite the applied force by the AFM.

The three experiments performed with different concentration of chloride and different pH levels are presented in Fig. 9. The graph shows the averaged topographical depth (nm) versus the time (s).

The solution with 0.01 M NaCl at neutral pH shows more loss in material than the one with 0.005 M NaCl at same pH which seems reasonable due to the different concentration of chlorides. The solution of 0.01 M NaCl in alkaline pH, does not present appreciable corrosion in agreement with the AFM observations of Fig. 5.

## 4 Conclusions

This paper presents experiments of simultaneous electrochemical measurements made in situ through an electrochemical cell specially adapted while obtaining topographical AFM images along the time.

According to the results, topographical images show that the ferrite phase corrodes preferentially in chloride containing solutions and agree well with averaged depth results calculated from LPR measurements and Faraday´s law,.

The LPR measurements show that in the initial states of corrosion, the Icorr increases quickly until the second hour. Afterwards, the mass loss first increases and then stabilizes at neutral pH experiment.

For both 0.01 and 0.005 M concentrations of NaCl, it is visible how ferrite dissolves acting as anode while the cementite acts as cathode.

We can observe how the surface does not change due to the passive layer formation in alkaline solution. The metal surface is protected from corrosion even in the presence of chloride ions.

## References


Andrade C (CSIC) (1989) Manual de inspección de obras dañadas por corrosión de armaduras. Madrid

Andrade C, Garcés P, Martínez I (2008) Galvanic currents and corrosion rates of reinforcements measured in cells simulating different pitting areas caused by chloride attack in sodium hydroxide. Corros Sci 50(10):2959–2964

Bertrand G (2000) In-situ electrochemical atomic force microscopy studies of aqueous corrosion and inhibition of copper. J Electroanal Chem 489(1–2):38–45

Cui GF (2006) A single precursor pit for pitting corrosion on defect of tinplate alloy layer visualized by atomic force microscopy. Mater Chem Phys 97(2–3):488–493

Datta J (2008) Role of Cl- and ions on the corrosion behaviour of 20% SiC reinforced 6061-Al metal matrix composite: a correlation between electrochemical studies and atomic force microscopy. Corros Sci 50(9):2658–2668

Galvele JR (1976) Transport processes and the mechanism of pitting of metals. J Electrochem Soc 123(4):464–474

Galvele J (1981) Transport processes in passivity breakdown-II. Full hydrolysis of the metal Ions. Corros Sci 21(8):551–579

Galvele JR (2005) Tafel's law in pitting corrosion and crevice corrosion susceptibility. Corros Sci 47(12):3053–3067

Gonzalez Fernandez JA (CSIC) (1989) Control de la corrosión. Estudio y medida por técnicas electroquímicas. Madrid

KamachiMudali U, Katada Y (2001) Electrochemical atomic force microscopic studies on passive films of nitrogen-bearing austenitic stainless steels. Electrochimica Acta 46(24–25):3735–3742
Li J, Meier DJ (1998) An AFM study of the properties of passive films on iron surfaces. J Electroanal Chem 454(1–2):53–58
Meira GR (2008) Modelling sea-salt transport and deposition in marine atmosphere zone – a tool for corrosion studies. Corros Sci 50(9):2724–2731
Nagarajan S, Rajendran N (2009) Crevice corrosion behaviour of superaustenitic stainless steels: dynamic electrochemical impedance spectroscopy and atomic force microscopy studies. Corros Sci 51(2):217–224
Nürnberger U (1984) Chloride corrosion of steel in concrete, part 2. Betonwerk Fertigteil-technik 10:697–704
Otero Huertas E (ed), Síntesis SA (2012) Corrosión y degradación de materiales. Madrid
Rongguang W (2004) An AFM and XPS study of corrosion caused by micro-liquid of dilute sulphuric acid on stainless steel. Appl Surf Sci 227(1–4):399–409
Sánchez J (2008) AFM study of the early corrosion of a high strength steel in a diluted sodium chloride solution. Corros Sci 50(7):1820–1824
Singh S (2009) Electrochemically controlled pitting corrosion in Ni film: a study of AFM and neutron reflectometry. Corros Sci 51(3):575–580
Wang R, Kido M (2006) Corrosion behaviours of pure iron beneath a micro-droplet of sulphuric acid solution investigated by atomic force microscopy. ScriptaMaterialia 55(7):633–636
Xu L-C, Chan K-Y, Fang HHP (2002) Application of atomic force microscopy in the study of microbiologically influenced corrosion. Mater Charact 48(2–3):195–203
Yanliang H, Kinsella B, Becker T (2008) Sensitisation identification of stainless steel to intergranular stress corrosion cracking by atomic force microscopy. Mater Lett 62(12–13):1863–1866
Zhang Q (2005) Observation by atomic force microscope of corrosion product during pitting corrosion on SUS304 stainless steel. ScriptaMaterialia 52(3):227–230

# 25 Years of Experience with Cathodic Protection of Steel in Concrete in the Netherlands

**Rob B. Polder and Willy H.A. Peelen**

**Abstract** This paper reviews 25 years of experience with cathodic protection of steel in concrete in The Netherlands. Three phases are distinguished from the late 1980s until present: pioneering, development and maturity. The application has grown considerably over these years, up to a total number of 150 structures with CP by the end of 2010. Technical and non-technical developments are highlighted and further growth is foreseen.

**Keywords** Cathodic protection • Reinforcement corrosion • Long term experience • Concrete

## 1 Introduction

Corrosion of reinforcing steel in concrete structures may occur, either due to penetration of chloride ions from de-icing salts or sea water spray or due to chloride ions having been mixed in as a set accelerator (Bertolini et al. 2013). Other causes of corrosion are carbonation of concrete, which only is relevant for concrete of poor quality or structures with low cover depths; or stray currents, which is limited to cases where concrete is subjected to strong direct current interference, mainly DC powered railways. Reinforcement corrosion causes concrete cracking, spalling of the cover and steel diameter reduction, eventually resulting in loss of safety. Conventional repair involves heavy, labour intensive and costly work. Competition pressure works against the required quality and quantity levels (cleaning of reinforcement, removal of contaminated concrete). Consequently, conventional repair is short lived in many cases. Corrosion reappears quickly and structures need to be repaired again after a relatively short time, further increasing their life-cycle cost. In a European study it was found that repairs had a short life in practice

R.B. Polder (✉)
Structural Reliability, TNO Technical Sciences, Delft, Netherlands

Civil Engineering and Geosciences, Delft University of Technology, Delft, Netherlands
e-mail: R.B.Polder@tudelft.nl

W.H.A. Peelen
Structural Reliability, TNO Technical Sciences, Delft, Netherlands



C. Andrade et al. (eds.), *Durability of Reinforced Concrete from Composition to Protection*, DOI 10.1007/978-3-319-09921-7_7

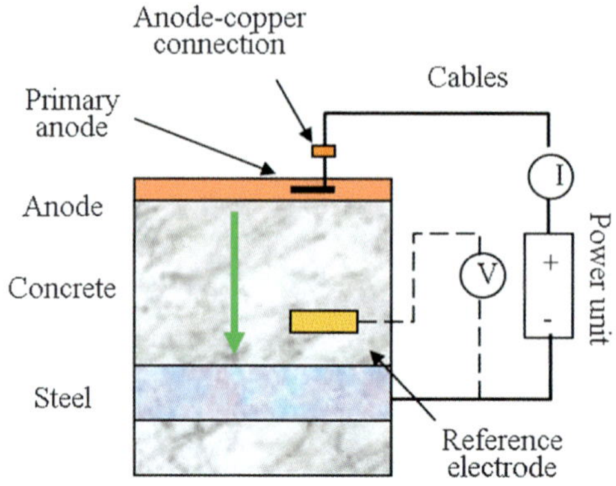


**Fig. 1** CP system principle layout and main components

(Tilly and Jacobs 2007) and a Dutch review of repair projects presented a similar result (Visser and van Zon 2012). A completely different situation exists for cathodic protection (CP) as a corrosion protection method. Cathodic protection of reinforcing steel has been applied to concrete structures with corrosion damage in Europe for over 25 years and slightly longer in the US (Grefstad 2005; Nerland et al. 2007; Polder 1998; Wenk and Oberhänsli 2007; Tinnea and Cryer 2008). A recent study reports on long term performance of CP systems in UK motorway structures (Christodoulou et al. 2010). This paper presents experience with CP of concrete in The Netherlands with reference to developments abroad.

Cathodic protection of steel in concrete works by supplying a small direct current from an external electrode, called the anode, to the reinforcing steel. This current polarises the steel into the negative direction, suppressing corrosion reactions and forcing (beneficial) cathodic reactions. The principle layout and the main components for a CP system with a conductive coating on the concrete surface are shown in Fig. 1.

## 2 Pioneering Phase 1987–1995

In the 1970s corrosion of steel in concrete developed on a large scale in the US. Early experiments with CP by Stratfull (Stratfull 1974) showed the effectiveness of the principle. In the mid-1980s American technology was introduced in Europe. The first CP system in The Netherlands was installed in 1986/1987 on a cycle lane of a bridge, the Stadium Viaduct in Rotterdam (Polder 1998). The anode was made up of carbon filled polymer cables with a copper wire core. This marked the start of the pioneering period of CP in The Netherlands. Subsequently, a few projects were made with this anode. Next, activated titanium anodes were introduced and used in a precast concrete façade. Titanium strips were placed in holes drilled longitudinally in 2,500 precast cantilever beams with mixed in chloride in two apartment buildings in 1990. It appeared that repair of precast concrete with mixed in chloride was a major problem area in the country, that was not solved effectively by conventional repair: within a few years corrosion and concrete cracking reappeared in many cases. On the other hand, CP showed to be effective and durable. Carbon based conductive coatings were introduced as anode material and applied to several buildings. Considering that uniform current distribution was important and would be mainly determined by electrical resistivity of the concrete, it was considered that the resistivity of materials used for repairs associated with CP should be similar to that of the parent concrete. CP companies and knowledge institutes worked together to draw up a CUR Recommendation published in 1996 (CUR 1996). It addressed curative application to reinforced concrete and had the character of a guideline including technical considerations and limiting values. In this period, a total of 16 structures had been provided with CP, mainly buildings with mixed in chloride, either with activated titanium anode strips in boreholes or with conductive coatings on the concrete surface.

## 3 Developmental Phase 1996–2004

In the next phase, application of CP to bridges and other infrastructure was developed. In one project, drilled in titanium strip anodes were foreseen in a thick wall of a basement of a bridge where corrosion induced delamination of the cover was widespread. However, reinforcing bars were not positioned at regular intervals as expected and many additional holes had to be drilled to avoid short circuiting between anode strips and bars. It also appeared impossible to install a regular maintenance contract between the managing organisation and the CP company. In another bridge corrosion had developed in the edge beam of a bridge deck above the abutments adjacent to leaking joints. The complication was that the deck was post-tensioned and concern existed for overprotection, causing hydrogen evolution and possibly hydrogen embrittlement of the prestressing steel. The problem was solved following a consult of Pietro Pedeferri, who made clear that monitoring the potential of the prestressing and maintaining it above a safe value would exclude hydrogen evolution (Pedeferri 1996). This approach and the safe value was later adopted in the European Standard EN 12696, that was published in 2000 (CEN 2000). In addition to allowing CP of prestressed concrete, it also included preventive application, Cathodic Prevention, which had been introduced by Pedeferri (Pedeferri 1996). The Standard had a performance based approach (rather than a technical guideline). In this period galvanic anodes were introduced for CP in concrete, based on the sacrificial action of zinc. Zinc sheet with an ion conductive adhesive was applied to both buildings and bridges. In this period, a total of 79 structures were provided with CP, comprised of buildings with mixed in chloride, bridges, parking structures and swimming pools. In some cases a single structure comprised more than one CP system, e.g. on the facades and on a basement floor.

# 4 Maturity Phase 2005–Present

In this period, CP has become a well-accepted method for long term protection of corroding structures. Involved companies, although relatively small and with economic ups and downs, joined in sponsoring a collective research project by TNO. It included setting up a database of CP systems in the country and further developing numerical modelling of CP (as follow up to previous work in collaboration with Politecnico Milano and in EU project ARCHES). These items will briefly be related here.

Ultimo 2010, a total of 150 concrete structures had CP installed, comprised of 105 buildings (dwellings, offices, parking and industrial buildings) and 45 bridges and other infrastructure, with a total of 85,000 square meters of protected concrete surface (Polder et al. 2012a). A diagram of the number of structures with a CP systems installed per year is shown in Fig. 2. These numbers do not include 65,000 $m^2$ of ground floor elements with mixed in chloride provided with CP.

Out of 150 CP systems, 65 had conductive coating anodes, 50 activated titanium and 35 galvanic anodes. For 105 good documentation was provided; 50 documented cases had been operating for 10 years or more and 55 cases for less than ten years. The remaining 45 cases were not documented and/or not monitored. The survey and subsequent analysis showed a mixed picture of some weak details and an overall low rate of system degradation. Weak details included accelerated corrosion of poorly insulated anode-copper connections; and some degradation of primary anodes in coating systems, in particular in older systems. Companies had improved most of these weak points, as was borne out by later systems. Very low numbers of overall anode degradation were found. However, local degradation of conductive coatings due to water leakage from overlying parts was an important cause of the need for (some) maintenance. In particular conductive coating anodes on buildings with relatively sheltered facades may last as long as 15–20 years. Activated titanium systems perform well over at least 20 years in documented cases; working lives of 25 and more years seem very well possible.

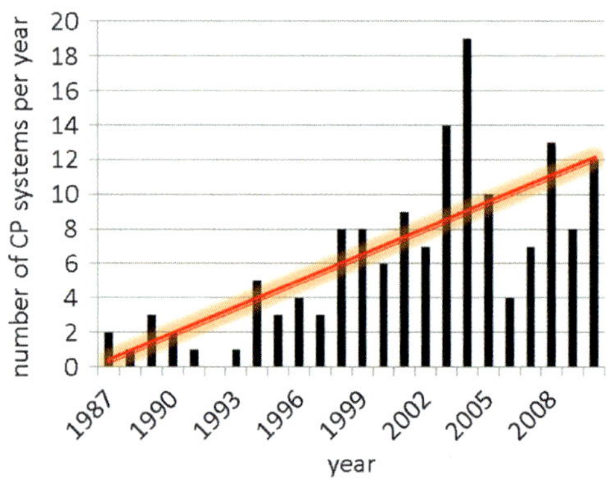


**Fig. 2** CP systems installed on concrete structures in The Netherlands per year (*bars*) and trendline

Numerical modelling of CP can be attractive in the design phase for minimising anode material while guaranteeing good protection where needed including in depth (throwing power). Such models are a combination of finite element modelling of steel and concrete geometry and electrochemical input parameters for steel, concrete and anode. First attempts were based on numerical modelling of corrosion (Redaelli et al. 2006) and showed a reasonable agreement with practical cases for preventive effects in a two-dimensional scheme (Polder et al. 2009). Recent work allows full three-dimensional modelling of sizeable elements (up to 1 $m^3$). However, further work is needed on time dependent effects, although their influence may be rather small beyond a few days or weeks (Polder et al. 2011). Importantly, the polarisation of prestressing steel can be predicted, which facilitates safe application to prestressed structures.

## 5 Changing External Conditions

The good experience with CP of concrete structures has been well known to a small group of scientists, engineers and companies involved in CP of concrete. It is becoming of much wider interest for the construction industry, however, as a consequence of the aging of concrete infrastructure. Aging means in many cases degradation in the form of reinforcement corrosion, in particular due to penetration of de-icing salts. From a survey of concrete motorway bridges in The Netherlands it has been found that at an age of 40 years about 5 % of these bridges have developed (chloride induced) corrosion related concrete damage, and 50 % has such damage after 70 years of service (Gaal 2004). The majority of Dutch concrete bridges has been built between 1960 and 1980. Combining with data on repair life mentioned above (Tilly and Jacobs 2007; Visser and van Zon 2012) and assuming a total number of all concrete bridges in the country of 40,000 (i.e. ten times the c. 4,000 motorway bridges), further analysis has predicted the amount of bridges with corrosion damage (Polder et al. 2012b). For a repair life of 10 years, the analysis shows that in 2012 about 500 bridges need repair; in 2024 some 1,000 bridges need repair and in 2040 about 2100. A significant (and increasing) number of these repairs involve repair of old repairs! These repairs would involve considerable costs and significant effects on traffic flow. Using CP with an assumed but not unrealistic mean life of 25 years, these numbers would reduce to roughly half. Consequently, large sums of money and societal impact can be saved by applying CP instead of conventional repair.

On another level, a significant development is that organisations responsible for infrastructure employ less and less technical experts. Risk and responsibility for maintaining infrastructure are more and more transferred to contracting companies, who traditionally have a focus on new construction. Contractors thus are facing responsibility for long-term maintenance of road networks including existing structures, for periods between 20 and 50 years. This poses challenges and opportunities: general contractors are usually not very experienced in concrete repair;

yet, a method like CP with a proven track record may very well fit in their business philosophy. Moreover, CP monitoring provides unequivocal data of good performance (absence of corrosion); which usually is the criterion for evaluation of such contracts.

In 2012 and 2013, CP is installed on 29 motorway bridges on a total of 1,300 corroding (prestressed) beam heads due to joint leakage. Maintenance is organised in a 20 year contract.

In some cases, owners have realised that corrosion can have expensive consequences. After a number of cases with corrosion induced damage due to leaking joints, the Province of Groningen now has decided on a pro-active policy. In the half-joints of a number of new bridges they install cathodic prevention (CPre). As shown by Pedeferri in the 1990s, a low current suffices to prevent corrosion, which allows economical systems to be installed with simple monitoring (Pedeferri 1996).

## 6 Conclusions

The experience of the last 25 years or so have shown that CP is a reliable and dependable method to halt corrosion of reinforcement in concrete structures. Lessons have been learned from previous projects, resulting in improved materials and details. Critical design issues can be addressed by numerical modelling. Overall confidence has grown with both CP companies and owners. Unit prices have decreased due to increased (long term) confidence and improved techniques. At the same time, prices for proper conventional repairs have increased. Based on the market in The Netherlands CP prices for protection of infrastructural works are competitive. CP systems work for a very long time and their good performance is systematically shown by measurements. It is expected that the market for CP of concrete will increase considerably in the next few years up to a decade, in particular in infrastructure.

## References

Bertolini L, Elsener B, Pedeferri P, Redaelli E, Polder RB (2013) Corrosion of steel in concrete: prevention, diagnosis, repair, 2nd edn. Wiley-VCH Verlag GmbH & Co. KGaA, Weinheim, 414 p. ISBN 3-527-33146-8

CEN (2000) Cathodic protection of steel in concrete. EN 12696

Christodoulou C, Glass GK, Webb J, Austin S, Goodier C (2010) Assessing the long term benefits of impressed current cathodic protection. Corros Sci 52:2671–2697

CUR, Gouda (1996) CUR Aanbeveling 45 Kathodische bescherming van wapening in betonconstructies, Cathodic protection of reinforcement in concrete structures, in Dutch

Gaal GCM (2004) Prediction of deterioration of concrete bridges, PhD thesis, Delft University Press, Delft

Grefstad K (2005) Cathodic protection applied on Norwegian concrete bridges. Experience and recommendations. Eurocorr05, Lisboa

Nerland OCN, Eri J, Grefstad KA, Vennesland Ø (2007) 18 years of cathodic protection of reinforced concrete structures in Norway – facts and figures from 162 installations. Eurocorr07, Freiburg

Pedeferri P (1996) Cathodic protection and cathodic prevention. Constr Build Mater 10(5):391–402

Polder RB (1998) Cathodic protection of reinforced concrete structures in the Netherlands – experience and developments. HERON 43(1):3–14

Polder RB, Peelen WHA, Lollini F, Redaelli E, Bertolini L (2009) Numerical design for cathodic protection systems for concrete. Mater Corros 60(2):130–136

Polder RB, Peelen WHA, Stoop B, Neeft E (2011) Early stage beneficial effects of cathodic protection in concrete structures. Mater Corros 62(2):105–110

Polder RB, Leegwater G, Worm D, Courage W (2012a) Service life and life cycle cost modelling of cathodic protection systems for concrete structures. Cem Concr Compos 47:69–74

Polder RB, Peelen WHA, Courage W (2012b) Non-traditional assessment and maintenance methods for aging concrete structures – technical and non-technical issues. Mater Corros 63(12):1147–1153

Redaelli E, Bertolini L, Peelen WHA, Polder RB (2006) FEM-models for the propagation of chloride induced reinforcement corrosion. Mater Corros 57(8):628–635

Stratfull RF (1974) Cathodic protection of a bridge deck. Mater Perform 13(4):24

Tilly GP, Jacobs J (2007) Concrete repairs – performance in service and current practice. IHS BRE Press, Bracknell

Tinnea JS, Cryer CB (2008) Corrosion control of pacific coast reinforced concrete structures: a summary of 25 years experience. In: The 1st international conference on heritage and construction in coastal and marine environment MEDACHS08, Portugal

Visser JHM, van Zon Q (2012) In: Alexander et al (eds) Performance and service life of repairs of concrete structures in The Netherlands, international conference on concrete repair, rehabilitation and retrofitting III. Taylor & Francis, London, ISBN: 978-0-415-89952-9

Wenk F, Oberhänsli D (2007) Long-term experience with cathodic protection of reinforced concrete structures. Eurocorr07, Freiburg

# Corrosion Induced Cracking of Reinforced Concrete

**Caroline Fahy, Peter Grassl, and Domenico Gallipoli**

**Abstract** In this work a coupled mechanical and transport lattice model based on Delaunay and Voronoi tessellations is used to model the penetration of corrosion products into the vicinity of the steel in a concrete specimen. The penetration of the corrosion products into the concrete is described by constitutive laws developed for volume transport through porous materials and the cracking resulting from this is modelled by the mechanical lattice using a damage mechanics constitutive law. The effect of penetration of the corrosion products into the pores and cracks on the time to cracking is investigated as part of this work. The numerical results are compared to experimental results reported in the literature.

**Keywords** Corrosion • Cracking • Lattice • Computational modelling

## 1 Introduction

Corrosion induced cracking and spalling has a major influence on the life of reinforced concrete structures. Therefore, it is desirable to develop models, which can predict this time dependent process. Corrosion induced cracking is caused by the constrained volume expansion of oxides produced by the corroding reinforcement bars. While the corrosion products are formed, they penetrate into the pores and micro-cracks of the concrete surrounding the reinforcement bar (Wong et al. 2010). In other researchers work, this penetration was captured by the inclusion of a porous zone surrounding the rebar or by allowing for cracks to be filled with rust (see for instance Liu and Weyers 1998; El Maaddawy and Soudki 2007; Chernin et al. 2010), without modelling explicitly the coupling of the mechanical and transport processes. For instance, for models including a porous zone, the mechanical pressure on the concrete was often only activated once the porous layer was assumed to be filled with rust. However, the transport of rust and

C. Fahy (✉) • P. Grassl
School of Engineering, University of Glasgow, Glasgow, UK
e-mail: c.fahy.1@research.gla.ac.uk; p.grassl@glasgow.ac.uk

D. Gallipoli
Laboratoire SIAME, Université de Pau et des Pays de l'Adour, Pau, France
e-mail: domenico.gallipoli@univ-pau.fr



C. Andrade et al. (eds.), *Durability of Reinforced Concrete from Composition to Protection*, DOI 10.1007/978-3-319-09921-7_8

the mechanical response are coupled. The pressure generated by the volume constraint of corrosion products results in the penetration of the products into the concrete. In this work, we aim to model this coupled process. The penetration is modelled by a volume transport lattice model for fully water saturated porous materials that is coupled to a mechanical lattice approach. It is assumed that the corrosion products act as a fluid permeating the porous concrete. The cracking of the concrete is modelled by a damage mechanics constitutive law and transport of the corrosion products is described by Darcy's law.

## 2 Lattice Model

In the present work, lattices based on one-dimensional mechanical and transport elements model the mechanical response and fluid transport. The specimen domain is discretised by dual Delaunay and Voronoi tessellations, which are shown in Fig. 1a. The mechanical elements are placed along the edges of the Delaunay triangles (Bolander and Berton 2004). The mid cross-sectional geometry of the elements is determined by the corresponding Voronoi polygon edges (Fig. 1b). The transport lattice is composed of one-dimensional transport elements placed along the edges of the Voronoi polygons (Grassl 2009). Their cross-sectional properties are determined from the corresponding edges of the Delaunay triangles. The coupling of the two models is achieved by an exchange of history variables at the point C in Fig. 1b, which contains information for both the mechanical and the transport models.

In Fig. 1b, $u$, $v$ and $\phi$ are the nodal degrees of freedom for the mechanical model.

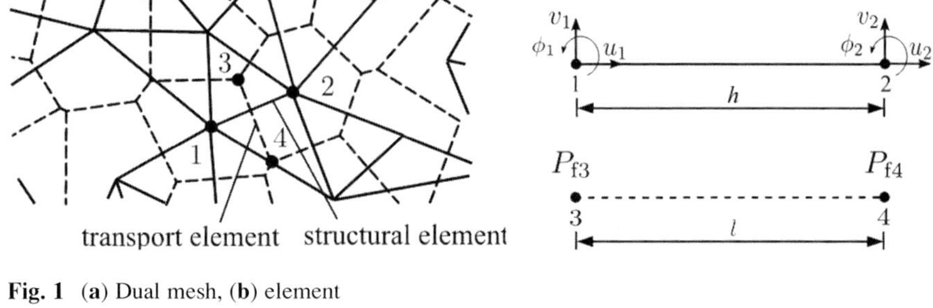


**Fig. 1** (**a**) Dual mesh, (**b**) element

## 2.1 Mechanical Model

An isotropic damage model that corresponds to a continuous reduction of the element's stiffness describes the inelastic response of the lattice element. The stress–strain law for this is

$$\boldsymbol{\sigma} = (1 - \omega)\mathbf{D} : \boldsymbol{\varepsilon} \tag{1}$$

where $\boldsymbol{\sigma} = \{\sigma_{\mathrm{n}}, \sigma_{\mathrm{s}}, \sigma_{\phi}\}^T$ is the stress vector (comprising of normal, shear and rotational strains), $\omega$ is a damage parameter, $D$ is the elastic stiffness and $\boldsymbol{\varepsilon} = \{\varepsilon_{\mathrm{n}}, \varepsilon_{\mathrm{s}}, \varepsilon_{\phi}\}^T$ is the strain vector (comprising of normal, shear and rotational strains). The elastic stiffness is defined as

$$\mathbf{D} = \begin{bmatrix} E & 0 & 0 \\ 0 & \gamma E & 0 \\ 0 & 0 & E \end{bmatrix} \tag{2}$$

and is dependent on the model parameters, $E$ and $\gamma$. For a plane stress analysis and a regular equilateral triangular lattice, these model parameters can be related to the continuum Young's modulus $E_{\mathrm{c}}$ and Poisson's ratio $\nu$ as follows

$$\gamma = \frac{1 - 3\nu}{\nu + 1} \tag{3}$$

$$E = \frac{E_{\mathrm{c}}}{1\text{-}\nu} \tag{4}$$

## 2.2 Transport Model

The volume balance equation presented in this section describes the stationary volume flow through a transport element subjected to fluid pressure gradient. It is assumed that the material is fully saturated. In the balance equation, the flux (i.e. the rate of liquid flow per unit area) is related to the pressure gradient through Darcy's law, which gives

$$k\,\mathrm{div}(\mathrm{grad}(P_{\mathrm{f}})) = 0 \tag{5}$$

where $k$ is the hydraulic conductivity and $P_{\mathrm{f}}$ is the fluid pressure. The discrete form of Eq. 5 is

$$\alpha \boldsymbol{P}_{\mathrm{f}} = \boldsymbol{f} \tag{6}$$

where $\alpha$ is the conductivity matrix and $\boldsymbol{f}$ is the nodal flow rate vector. The conductivity matrix is defined as

$$\alpha = \frac{Ak}{l}\begin{pmatrix} 1 & -1 \\ -1 & 1 \end{pmatrix} \tag{7}$$

in which $A$ is the cross-sectional area and $l$ is the length of the transport element. The hydraulic conductivity $k$ consists of

$$k = k_o + k_c \tag{8}$$

in which $k_o$ is the hydraulic conductivity of the uncracked material and $k_c$ is the additional conductivity due to cracking. The hydraulic conductivity $k_o$ can be expressed in terms of absolute (dynamic) viscosity $\mu$ and intrinsic conductivity $\kappa$ as

$$k_o = \frac{\kappa}{\mu} \tag{9}$$

The influence of different values of dynamic viscosity $\mu$ on the time to cracking is investigated in Sect. 3.2. The cracking part of the conductivity $k_c$ is

$$k_c = \xi \frac{\overline{\omega}_c^3}{12\mu h} \tag{10}$$

where $\xi$ is a factor taking into account the tortuosity of the crack surface, $\overline{\omega}_c^3$ is the cube of the equivalent crack opening of the mechanical lattice and $h$ is the length of the mechanical element. A number of different values of the tortuosity factor are considered in this work to investigate its effect on the time to cracking (Sect. 3.3).

The more permeable the concrete is, the more corrosion products are transported into the concrete and the longer it takes for the concrete to crack.

## 3 Analysis of Corrosion Induced Cracking

In the present section, the proposed coupled lattice approach is applied to the plane stress analysis of a concrete specimen with a single eccentrically placed reinforcement bar. The specimen geometry was taken from the experimental study on corrosion induced cracking presented by Andrade et al. (1993) and is shown in Fig. 2. Table 1 contains the model parameters used for the analyses. In the analyses, the crack opening was calculated by measuring the relative displacement between two nodes placed along the top of the specimen (indicated by two circular dots in Fig. 2).

For the mechanical model, the complete specimen was meshed including the rebar. The model properties were specified for the concrete, rebar and the interface elements where the corrosion products form. The interface elements were given the same material properties as the concrete elements, except the shear stiffness of these elements, which is chosen to be considerably smaller (Table 1) to allow cracks in the vicinity of the interface to open. As the rebar is assumed to be impermeable to fluid, the rebar was not meshed for the transport model.

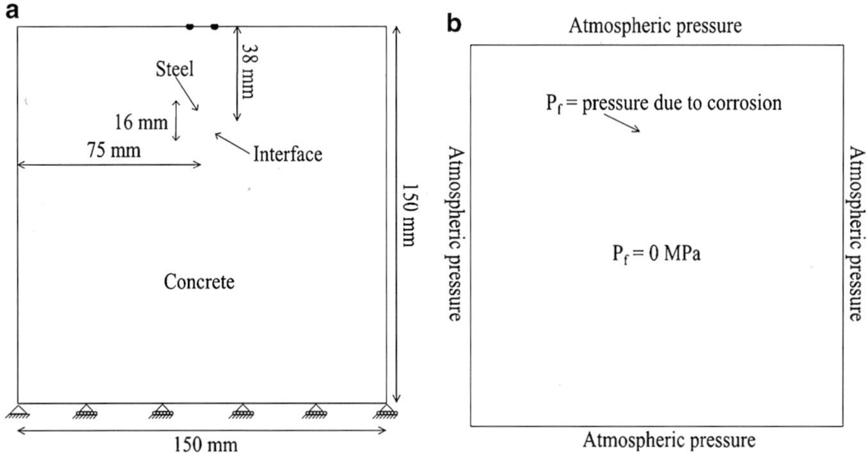


**Fig. 2** Concrete specimen 3 according to Andrade et al. (1993): (**a**) Mechanical geometry and setup, (**b**) transport boundary conditions

**Table 1** Material parameters

| | Concrete | Steel | Interface | Corrosion products |
|---|---|---|---|---|
| $f_t$ | 3.195 MPa | – | 3.195 MPa | – |
| $E$ | 36.995 GPa | 262.5 GPa | 36.995 GPa | – |
| $\gamma$ | 0.3333 | 0.3333 | $1.0 \times 10^{-6}$ | – |
| $\kappa$ | $1.0 \times 10^{-13}$ mm$^2$ | – | $1.0 \times 10^{-13}$ mm$^2$ | – |

## 3.1 Application to Corrosion Products

In this work, the formation of corrosion products is modelled by applying an Eigen-displacement to the mechanical elements along the rebar/concrete interface. In Fig. 3a, the thick black line indicates the edge of the rebar, the grey solid lines are mechanical elements and the transport elements are shown as broken lines. The mechanical elements inside the rebar are not shown. Each mechanical element that crosses the edge of the rebar has an Eigen-displacement applied to it.

The applied Eigen-displacement results in a mechanical stress $\sigma$, which is used as the boundary condition for the transport model assuming that a layer of corrosion products separates the rebar and the concrete. Using the following equation (Eq. 11), the mechanical stress can be transformed to a fluid pressure (Fig. 3b)

$$P_\mathrm{f} = \frac{\sigma_1 l_1 + \sigma_2 l_2}{l_1 + l_2} \tag{11}$$

This fluid pressure results in the penetration of corrosion products into the concrete.

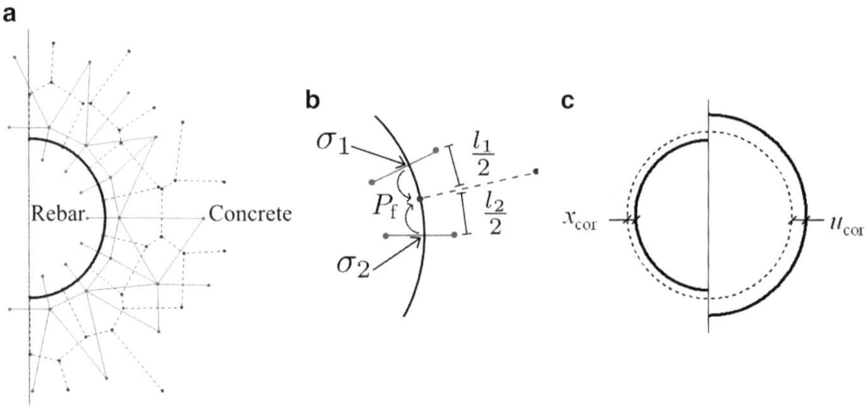


**Fig. 3** (**a**) Illustration of a typical meshed region surrounding the rebar, (**b**) close-up detail showing the application of fluid pressure due to the formation of corrosion products, (**c**) attack penetration

The total amount of corrosion products generated can be related to the attack penetration depth by carrying out the following steps. Firstly, an Eigen-displacement is applied to the mechanical elements on the interface between the rebar and concrete, which results in compressive stresses in these elements. The compressive stresses are used to calculate the fluid pressure, which drives the transport model causing the corrosion products to penetrate through the concrete. The volume of applied Eigen-displacement is added to the volume of rust flowing into the pores in the transport model. Converting this total volume of corrosion to a length gives $u_{cor}$ as shown in Fig. 3c. To determine the corresponding attack penetration depth $x_{cor}$ (Fig. 3c), $u_{cor}$ is divided by an assumed expansion factor, which is chosen as 2 as used by Molina et al. (1993).

## 3.2 *Investigation of the Effect of Dynamic Viscosity on the Time to Cracking*

Little information is known about the properties of corrosion products. In this work, the rust is assumed to behave as a fluid. An investigation of the effect of different values of viscosity on the time to cracking was carried out and in this case the effect of cracking on the conductivity was ignored ($\xi = 0$). Figure 4 shows the crack widths obtained against the corresponding attack penetration depth when different values of dynamic viscosity are assumed. The dimensionless parameter $\overline{\mu}$ is the ratio between the value of dynamic viscosity assumed and that of water.

It can be seen that the dynamic viscosity of the fluid has a strong influence on the crack width. Some researchers have suggested that corrosion products could be given similar mechanical material properties to that of water, as this is one of the main constituents (Molina et al. 1993). Expanding this assumption to the fluid properties, the normalised dynamic viscosity would be 1. However, this does not

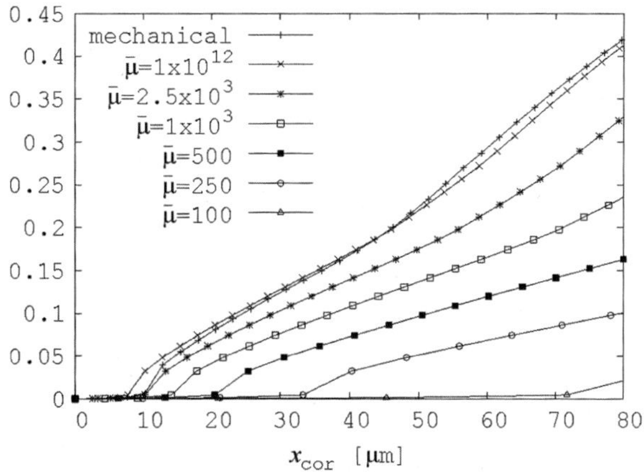


**Fig. 4** Effect of dynamic viscosity on crack widths $\xi = 0$

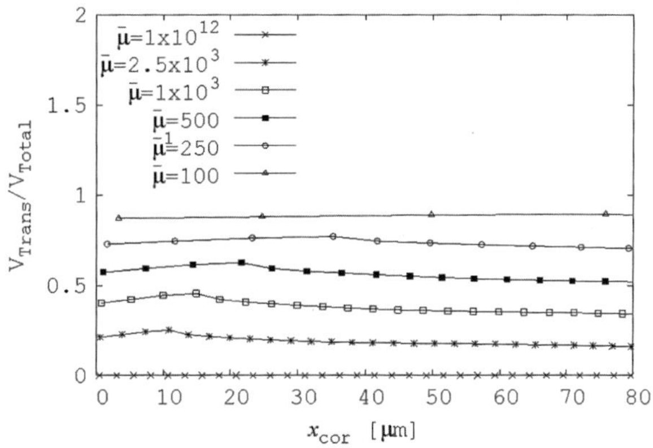


**Fig. 5** Ratio of volume of corrosion products transported into pores and total volume of corrosion products versus penetration depth for different viscosity values

lead to the initiation of cracking in the specimen within the expected time period, as shown in Fig. 4. Furthermore, if the value of dynamic viscosity chosen is very large, then the pressure required for the corrosion products to penetrate the pores and micro-cracks is so great that cracks are induced at a similar time to when a purely mechanical model is used.

Figure 5 shows the ratio between the volume of corrosion products penetrating the pores and the total volume of corrosion products being produced. With decreasing viscosity values, the ratio increases. This is expected, as the corrosion products with a lower viscosity can penetrate the pores easier than those with a higher viscosity.

### *3.3 Investigation of the Effect of Tortuosity on the Time to Cracking*

The formation of cracks in the specimen causes a change in the hydraulic conductivity of the elements. For these analyses, the value of the dynamic viscosity was chosen as $\overline{\mu} = 1.0 \times 10^3$ based on the findings of the investigation discussed in Sect. 3.2. The influence of the tortuosity on the conductivity is reflected in Eq. 10. One of the factors influencing the change in conductivity is $\xi$ and takes into account the roughness of the crack surface. The effect of this parameter on the time to cracking was also studied as part of this work. A $\xi$ value of 0 was initially chosen; this corresponds to the conductivity remaining unchanged despite the development of cracks. Three alternative values of $\xi$ were also considered; $1.0 \times 10^{-7}$, $1.0 \times 10^{-6}$ and $1.0 \times 10^{-5}$. Figure 6 shows the influence of the tortuosity factor on the time to cracking. It can be seen that the tortuosity factor has a strong influence on the crack widths obtained when the same amount of corrosion products are applied. As the tortuosity factor value is increased, the crack width obtained for a specific amount of applied corrosion products is reduced.

In Fig. 7, the ratio between the volume of corrosion products penetrating the pores and cracks surrounding the rebar and the total volume of corrosion products being produced are shown for different tortuosity values. An increase of the tortuosity factor results in an increase of the amount of corrosion products being transported in the concrete.

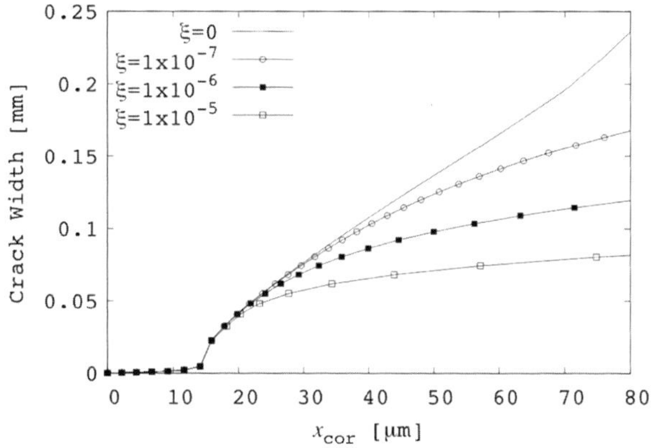


**Fig. 6** Effect of tortuosity on crack widths when $\overline{\mu} = 1.0 \times 10^3$

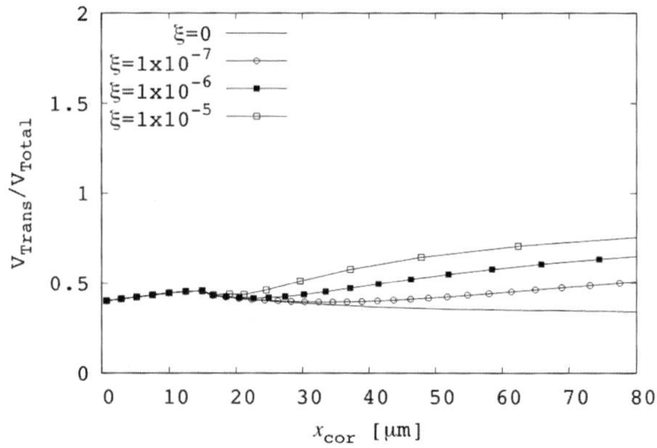


**Fig. 7** Ratio of volume of corrosion products transported into pores and total volume of corrosion products versus penetration depth for different tortuosity factors

**Fig. 8** Comparison of numerical and experimental results for specimen 3 by Andrade et al. (1993) when $\bar{\mu} = 1.0 \times 10^3$ and $\xi = 1.0 \times 10^{-7}$

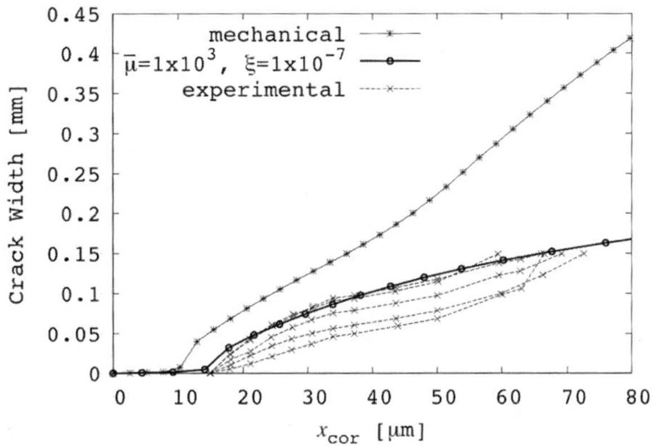


## 4 Comparison of Numerical and Experimental Results

Using the best-fit parameters for the viscosity and the tortuosity, determined during the investigations carried out in Sects. 3.2 and 3.3, the numerical results obtained were compared to experimental ones reported by Andrade et al. (1993). Figure 8 shows the comparison between a purely mechanical model, the proposed coupled model and the experimental results. The mechanical model (no penetration of corrosion products allowed) under-predicts the time to cracking. However, good agreement between numerical results and the experimental data is obtained when the effect of the penetration of the corrosion products is considered.

## 5 Conclusions and Future Work

A lattice approach that couples mechanical and volume transport has been applied to study the effect of the penetration of corrosion products into pores and micro-cracks. An investigation into the influence of the penetration of the corrosion products into the pores and micro-cracks was carried out in this work. The amount of corrosion products penetrating the pores relative to the total amount produced was presented. This study has shown that the time to cracking is strongly dependent on both the dynamic viscosity of the fluid and the tortuosity factor.

Comparison of a purely mechanical model with an expansion coefficient of 2, the proposed coupled model and experimental data showed the importance of allowing for the penetration of the rust into the pores. The numerical results obtained from the proposed coupled model are in much better agreement with the experimental results than when a purely mechanical model is used.

In future work, the effect of the pressure generated by the corrosion products on the mechanical stresses will be investigated and a time-dependent stiffness of the corrosion products will be included as part of the proposed model. Using the best fit parameters determined in this work and in future research, the proposed coupled model will then be applied to additional specimens and the numerical results obtained will be compared to those presented in the literature. This will assess the model's ability to predict the time to cracking of corroded reinforced concrete.

**Acknowledgement** The numerical analyses were performed with the nonlinear analyses program OOFEM (Patzak and Bittnar 2001) extended by the present authors.

## References

Andrade C, Alonso C, Molina F (1993) Cover cracking as a function of rebar corrosion: part I – experimental test. Mater Struct 26:453–464

Bolander J, Berton S (2004) Simulation of shrinkage induced cracking in cement composite overlays. Cem Concr Compos 26:861–871

Chernin L, Val L, Volokh K (2010) Analytical modelling of concrete cover cracking caused by corrosion of reinforcement. Mater Struct 43:543–556

El Maaddawy T, Soudki K (2007) A model for prediction of time from corrosion initiation to corrosion cracking. Cem Concr Compos 29(3):168–175

Grassl P (2009) A lattice approach to model flow in cracked concrete. Cem Concr Compos 31:454–460

Liu Y, Weyers R (1998) Modeling the time-to-corrosion cracking in chloride contaminated reinforced concrete structures. Mater J 95:675–681

Molina F, Alonso C, Andrade C (1993) Cover cracking as a function of rebar corrosion: part 2 – numerical model. Mater Struct 26:532–548

Patzak B, Bittnar Z (2001) Design of object orientated finite element code. Adv Eng Softw 32:759–767

Wong H, Zhao YX, Karimi AR, Buenfeld NR, Jin WL (2010) On the penetration of corrosion products from reinforcing steel into concrete due to chloride-induced corrosion. Corros Sci 52:2569–2480

# Influence of Crack Width on Long Term Degradation of Concrete Structures

**Julio Torres and Carmen Andrade**

**Abstract** The study of the influence of concrete crack width on corrosion of steel reinforcement due to exposure to aggressive environments is very important for durability design. Premature failure of structures may eventually occur during their service life. Since cracks in the concrete cover are inevitable due to the stress structures are subjected to, penetration of aggressive substances will occur more rapidly in cracked areas than in areas without cracking. This study addresses the corrosion effects on the reinforcement in cracked areas. Reinforced beams 2 m long and cracked at half-length with widths up to 1 mm were prepared 20 years ago, exposed to carbonation and chlorides and kept outdoors sheltered from rain until present. After 20 years, the carbonation depth in the faces of the cracks, and the reinforcement at the crack region have been examined. The results indicated that flexural cracks in high quality carbonated concrete present only a small additional risk of corrosion which, however, is significant in the presence of chlorides.

**Keywords** Crack • Durability • Concrete structures

## 1 Introduction

Durability of concrete structures is usually studied in untracked conditions, despite the fact that codes recognize the risk that a crack represents for the service life of the structures, because of the easier arrival of aggressive substances to the reinforcement.

Cracked areas may present zones of localized corrosion. The oxides that are generated can seal the crack, especially if it has a small width, stopping or lowering the corrosion process, or they can stress the surrounding concrete and enlarge the crack width, increasing the damage progressively. The belief that larger crack widths enable higher corrosion damage has led the codes to set limits to cracking induced tensile loads, considering the reinforcing and prestressing steel and the environment exposure class.

J. Torres (✉) • C. Andrade
Department of Durability, (IETcc-CSIC), Institute of Construction Sciences Eduardo Torroja, Madrid, Spain
e-mail: juliotorres@ietcc.csic.es

 
C. Andrade et al. (eds.), *Durability of Reinforced Concrete from Composition to Protection*, DOI 10.1007/978-3-319-09921-7_9

However, from studies of around 25 years ago by Beeby (1978, 1983), Nürnberger (1984), Schiessl (1988), Arya and Ofari-Darko (1996), it was concluded that although the crack is a risk, their width cannot be directly related to the corrosion severity developed. Parameters such as water/cement ratio or the cover thickness may be more decisive than the crack width.

According to more recent works by Tarek et al. (2001) and Otieno et al (2010), cracks not exceeding 0.4 mm, even if they involve a shorter path to aggressive penetration, especially chlorides, in terms of life do not appear to present a significant corrosion risk.

This paper summarizes a long term study on cracked beams exposed outdoor sheltered from rain. It also contains data in which the aggressiveness has been artificially increased by watering the cracked regions with tap water or a chloride salt solution during some months.

## 2 Experimental Procedure

In 1992, 34 prismatic concrete beams of 8 × 9 × 190 cm in size were fabricated (Fig. 1), which were reinforced with a single corrugated bar of 10 mm diameter. In 18 beams the bars were unprotected and in 16 beams they were galvanized. The cover depths selected were 20 and 40 mm. Two different cements type CEM I were used: a low alkali content and high alkali content type CEMI-45A, they were named: Atlántico and Asland. The proportions of the concrete mix are given in Table 1. The concrete was made in a manual concrete mixer. The curing was in wet conditions for 28 days.

**Fig. 1** Beams and crack zone

**Table 1** Composition of concrete

| Cement kg/m$^3$ | Gravel (6–12) kg/m$^3$ | Sand kg/m$^3$ | Admixture sikament cm$^3$/l | Water l | W/C |
|---|---|---|---|---|---|
| 400 | 1,160 | 840 | 19 | 150 | 0.375 |

## 2.1 Crack Formation

In order to induce flexural cracks perpendicular to the main reinforcement direction, after being cured two beams were placed face-to-face. A cylinder was introduced in the center between them and the beams were loaded by means of a metallic wire attached to the protruded parts of the reinforcing bars on both sides until certain crack widths were achieved in the loaded central zone. The crack widths obtained ranged between 0.1 and 0.6 mm. Several secondary cracks of lower widths were also produced in some beams (Fig. 1).

## 2.2 Exposure Regime

### 2.2.1 First Period

The beams after fabrication and curing were submitted to different aggressive regimes in order to accelerate the possible corrosion process (Sarría Odiaga 1997). The sequence of exposure conditions is summarized in Table 2. The stages and the exposure regimes have been numbered for the sake of illustration. After the stage 1 and 2, five beams (see Table 3) were selected for exposure to a chloride solution during 12 months. The cores drilled at the end of stage 3 (1,985 days),

**Table 2** Summary of exposure periods and stages of the cracked beams

| | | | Partial (days) | Total (days) |
|---|---|---|---|---|
| Stage 1 (34 beams) | | Manufacture | 1 | 117 |
| | | Curing (underwater) | 28 | |
| | | Carbonation (sheltered from rain) | 28 | |
| | | Wet-dry cycles | 60 | |
| STAGE 2 (34 beams) | | Carbonatatión (sheltered from rain) | 1,469 | 1,469 |
| STAGE 3 (5 beams) | Preliminary | Measuring time | 60 | 399 |
| | Salt solution and drying cycles | 1° Salt application | 42 | |
| | | 1° Dry cycle | 21 | |
| | | 2° Salt application | 42 | |
| | | 2° Dry cycle | 21 | |
| | | 3° Salt application | 42 | |
| | | 3° Dry cycle | 21 | |
| | Salt solution continuous | Dry cycle | 30 | |
| | | Salt application | 60 | |
| | | Dry cycle | 50 | |
| | | | | 1,985 |

**Table 3** Beams exposed to salt/dry cycles (stage 3)

| Beam | ASL9 | ASL 15 | ATL7G | ASL11G | NITALT 1 |
|---|---|---|---|---|---|
| Manufacturing year | 1992 | 1992 | 1992 | 1992 | 1992 |
| Cement type | Asland | Asland | Atlantico | Asland | Atlantico |
| Steel | Normal | Normal | Galvanized | Galvanized | Normal |
| Inhibitor | No | No | No | No | Nitrites |
| Crack size (mm) | 0.3 | 0.1 | 0.2 | 0.1 | 0.4 |
| Covering (mm) | 40 | 20 | 20 | 20 | 40 |

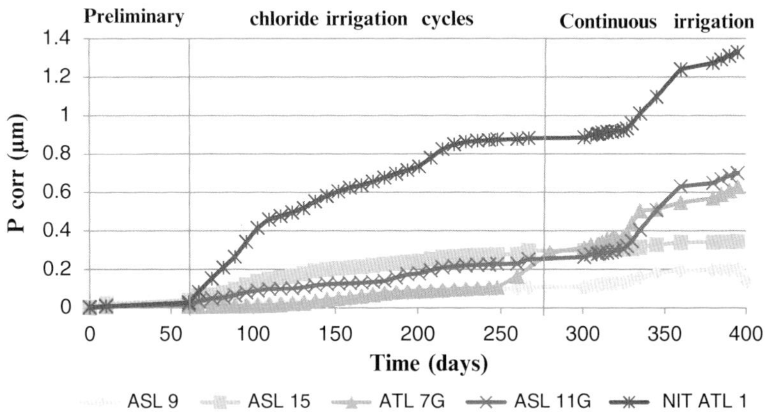


**Fig. 2** Accumulated corrosion (corrosion depth) during stage 3

indicated that the cracked zone was not carbonated at the bar position (of 20 and 40 mm cover).

The accumulated corrosion values $P_{corr}$ are given in Fig. 2. It can be deduced that in carbonated concrete in contact with salt solution the corrosion increases when the steel comes into contact with the chlorides, with the beam containing nitrites showing the highest value.

It can be deduced that depending on the beam type the $P_{corr}$ ranges from 0.2 to 1.4 μm (2 to 14 μm if localized attack is considered) which taking into account the short time (about 1 year test) can be classified as moderate active corrosion. Regarding the irrigation regime, the continuous watering seems to induce higher accumulated corrosion than cyclic irrigation, which enables to deduce that the wetness time is a key parameter. With respect to the type of steel, the beam with nitrites presented the highest values followed by the galvanized steel.

Representing the average value of corrosion rate in this stage 3 in function of the crack widths (Fig. 3), the values indicate that the corrosion rate with crack widths of 0.4 mm, is higher than with smaller sizes. Because the cracked zone was not carbonated at rebar level, the corrosion is attributed to the chlorides only. To confirm their arrival, chlorides were analyzed in two of the beams near the steel/crack interface obtaining values between 1.46 and 1.84 % by cement weight, which indicates an amount above the threshold value for active corrosion.

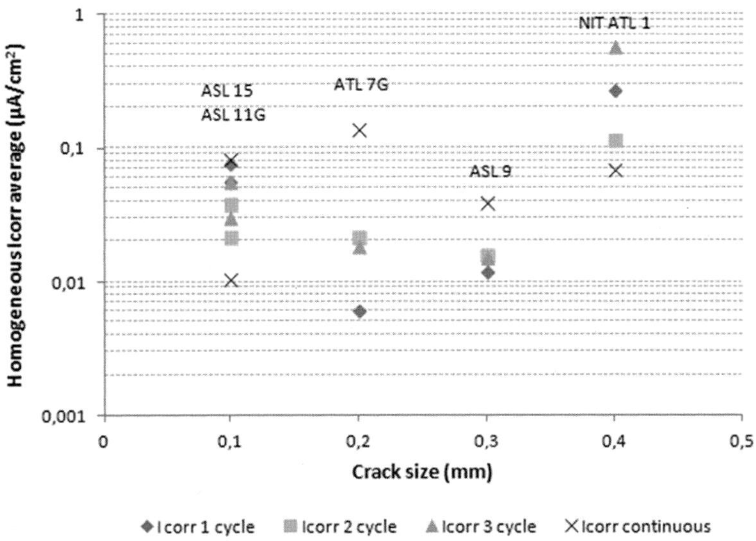


**Fig. 3** Average homogeneous $I_{corr}$ in function of crack width during stage 3

### 2.2.2 Second Period (Stage 4)

The experiment was stopped in 1992 after this first period but the beams were held at the atmosphere sheltered from rain until that in 2012 it was decided that the experiment should be repeated on 18 beams. It was found that the crack widths have changed a bit being now between 0.05 until 0.7 mm, while initially the range was 0.1–0.6 mm.

Corrosion measurements were made in all beams to characterize the state after this 20 years period.

After the corrosion measurements, four beams (Table 4) were selected to be characterized by determination of carbonation depth, porosity, and visual examination of rebar in the cracked zone. On four other beams (Table 4) corrosion measurements were performed during 4, 7 and 12 months after contact with tap water and chloride solution.

## 3 Results

When measuring after the 20 years period, as the beams collected were those initially only carbonated and they were held in sheltered conditions, the corrosion rate indicated that for all beams the embedded reinforcements were in a passive state, except one beam that showed higher values of the corrosion rate but below the

**Table 4** Beams tested in second period

| Stage 4 from 2012 | | | | |
|---|---|---|---|---|
| Name | Steel | Cover (cm) | Crack size (mm) | Measurement |
| 1A | Normal | 4 | 0.7 | Carbonation depth |
| 2AG | Galvanized | 4 | 0.5 | Porosity |
| 7AG | Galvanized | 2 | Negligible | Visual rebar |
| 8A | Normal | 2 | 0.08 | |
| 1B | Normal | 4 | 0.7 | Corrosion rate |
| 2BG | Galvanized | 4 | 0.2 | Corrosion potential |
| 7BG | Galvanized | 2 | 0.2 | Resistivity |
| 8B | Normal | 2 | 0.3 | |

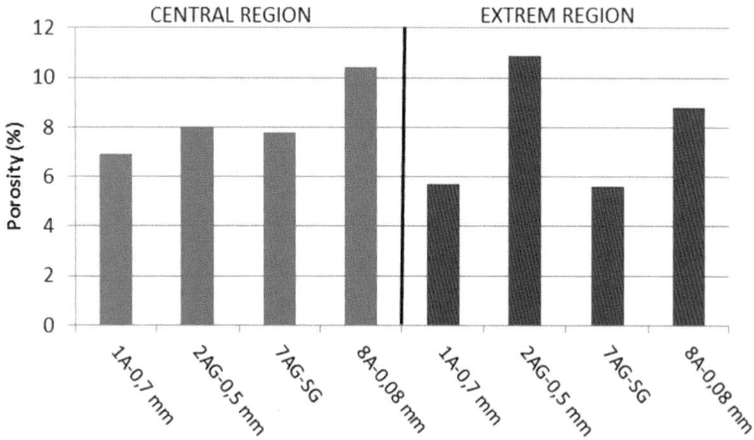


**Fig. 4** Porosity values of beams stage 4

threshold of 0.1 μA/cm$^2$. In accordance, the resistivity values were high, around 1,000 kΩcm, and the corrosion potentials were in general in the positive range. The mercury porosimetry values showed that the center of the beam is more porous than the sides, except in beam 2AG-0.5 mm. These values should have been all very similar as they were fabricated from the same batch (Fig. 4).

In Fig. 5 and the Table 5 the carbonation results are shown after this long exposure period from 1992. The crack faces and also the steel concrete interface in all beams are fully carbonated, except in the beam 2AG, that has a steel/concrete interface not carbonated. The beams 1A and 8A with normal steel presented a pit just in the cracked zone, while the beams 2AG and 7AG with galvanized rebars, appeared completely free of corrosion. It is important to note that in uncracked zones the carbonation had not reached the reinforcement after 20 years of exposure.

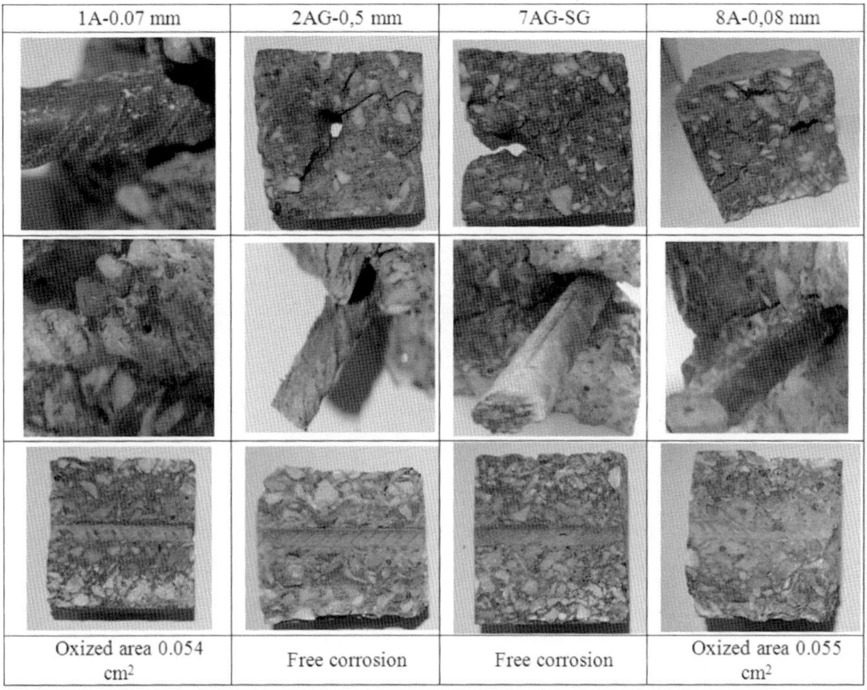


**Fig. 5** carbonated results and localized attack

**Table 5** Summary of observations testing after >20 years exposure sheltered from rain

| Beam | Region | steel | State crack face | State steel | State "back bar face" | Carbonation depth (mm) | |
|---|---|---|---|---|---|---|---|
| | | | | | | Top surface | Bottom surface |
| 1A-0.7 mm | Crack | N | Carbonated | Localized corrosion | Carbonated | 35 | 29 |
| | No crack | N | – | No corrosion | No carbonated | 5.1 | 10.7 |
| 2AG-0.5 mm | Crack | G | Carbonated | No corrosion | No carbonated | 35.1 | 31.1 |
| | No crack | G | – | No corrosion | No carbonated | 16.8 | 15.9 |
| 7AG-SG | Crack | G | Carbonated | No corrosion | Carbonated | 18.5 | 12.1 |
| | No crack | G | – | No corrosion | No carbonated | 11.1 | 7.2 |
| 8A-0.08 mm | Crack | N | Carbonated | Localized corrosion | Carbonated | 25.2 | 22.5 |
| | No crack | N | – | No corrosion | No carbonated | 8.8 | 8.6 |

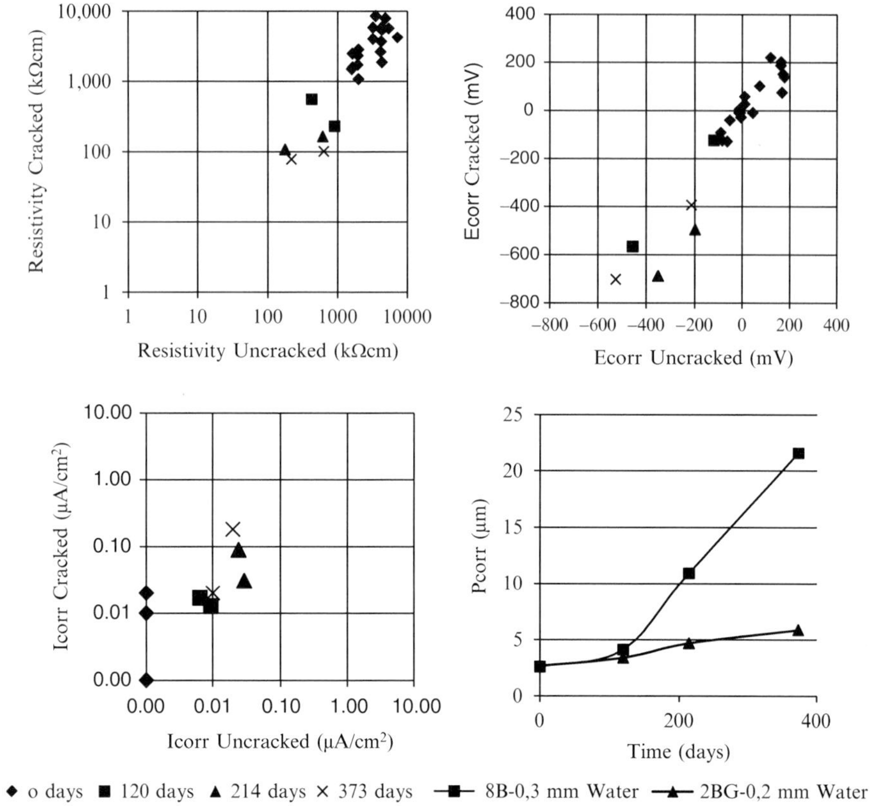


**Fig. 6** Resistivity, $E_{corr}$, $I_{corr}$ and $P_{corr}$ in tap water

In the tests made after the beams were exposed to water or chloride solution, corrosion tests were performed in only four beams in which no cores were drilled. Those in tap water presented both in cracked and uncracked zones an increase of corrosion rate with a simultaneous decrease of the resistivity and a shift to a more active (negative) level of the potentials. However, the changes are higher in the cracked zone. This is shown in Fig. 6 by comparing the values of the corrosion parameters in the cracked with the uncraked zones. The most significant change is noticed at 373 days (marked by crosses in Fig. 6). Regarding the crack width influence, the accumulated corrosion, $P_{corr}$ (Eq. 1), measured is higher in the beam with higher crack width (Fig. 6).

The beams with chloride solution depicted by Fig. 7 show the same trend but with even larger shifts of the corrosion parameters in the cracked zones and much higher (absolute) corrosion rates.

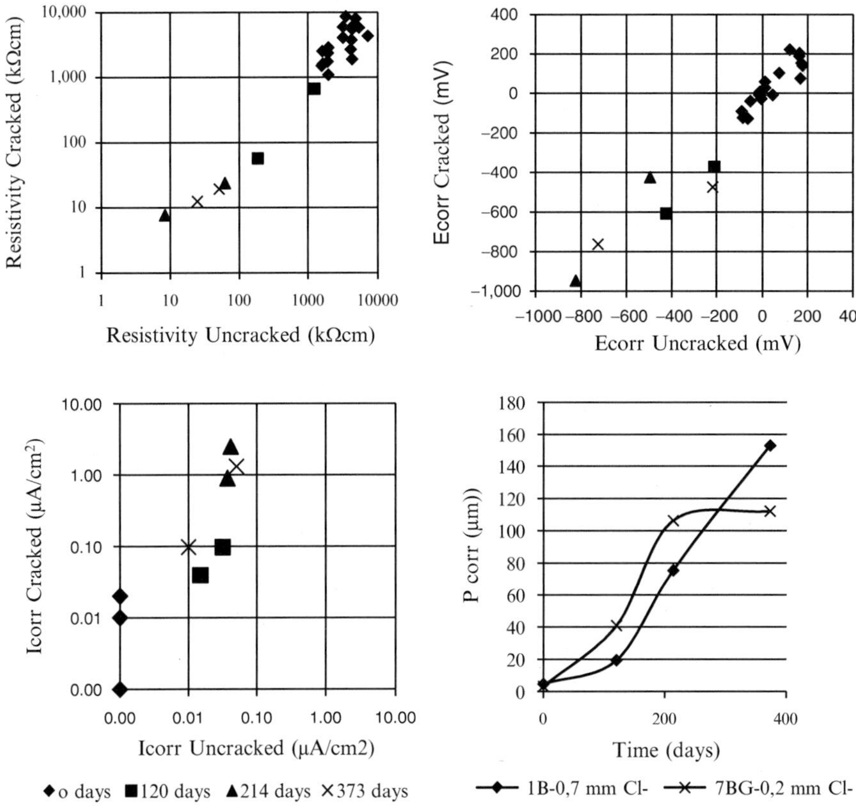


**Fig. 7** Resistivity, $E_{corr}$, $I_{corr}$ and $P_{corr}$ in chloride solution

# 4 Discussion

The corrosion attack in dry environment due to carbonation after around 20 years of exposure is negligible. Carbonation in uncracked areas has been partial and in no case has reached the reinforcement. However, cracked areas have corrosion on bare steel bars in the form of small pitting in the contact zone between the crack and reinforcement. After 20 years of exposure sheltered from rain, it is observed that the faces of the cracks are fully carbonated and carbonation has reached the bar position. After the period of 20 years it can be concluded that in spite of the low damage, cracks sizes greater than 0.08 mm, increase the depth of carbonation, thus facilitating pitting in the area of contact between crack and reinforcement. Wetting of the cracked zones with chlorides solution significantly increases the corrosion observed.

In stage 4 of the testing, related to recent measurements, it is confirmed, that also the bars were activated in the crack zone when simply watered with tap water (even without chlorides). With respect to the crack size, in Fig. 8 the relation is

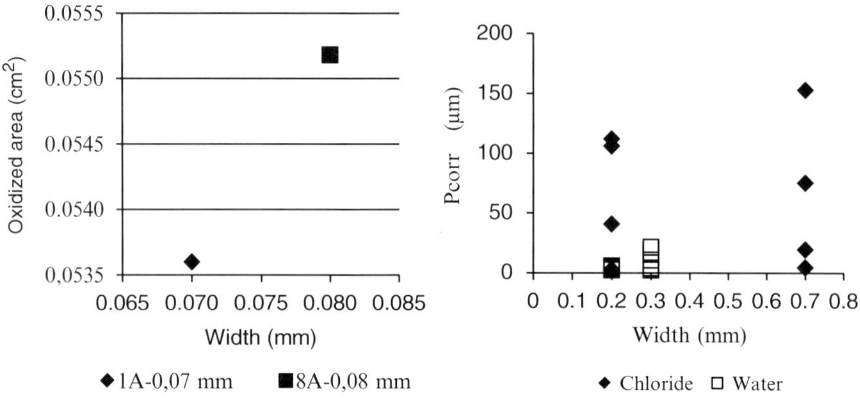


**Fig. 8** Relation between oxidized area and crack width, and dependence of accumulated corrosion $P_{corr}$ with crack width

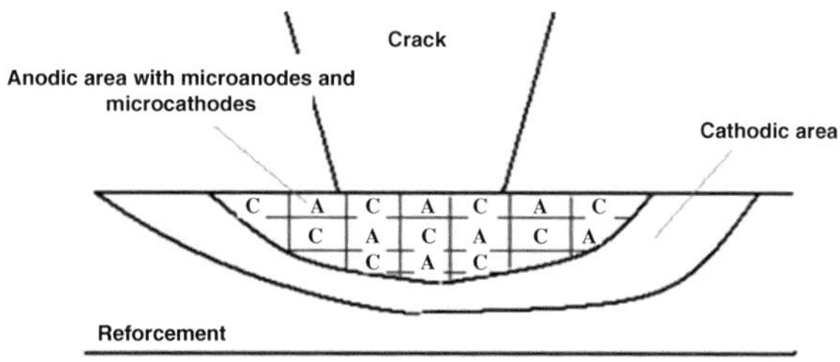


**Fig. 9** Macrocell in corroding area

summarized between corroded area and crack width and that of corrosion penetration depth and crack width. Although there is an apparent higher value for wider cracks, the scatter does not allow a definitive conclusion to be drawn.

With respect to the mechanism of corrosion in the cracked zone, the corroded area in itself is interpreted as a macrocell (which in its own is composed of anode + cathode microcells), which in turn is the anodic zone of a pure cathodic zone surrounding the corroded zone (Fig. 9).

Finally, Fig. 10 depicts the aspect of actual local attack (with and without the rust) produced in the cracked zone of one of the 34 beams (see Table 2) having been submitted to stage 1 (cycles of wet-dry during 60 days). The actual corrosion penetration depth found was about 215 µm. As no chlorides were applied to this beam, this attack should be interpreted to be due to carbonation from the end of the curing, being the active corrosion attributed to the wet cycles. Then, a corrosion of 215 µm during 60 days results acorrosion rate of 1,308 µm/year (around 113 $\mu A/cm^2$). As it is localized it can be divided by the pitting factor $\alpha = 10$ which ends in assuming that, if the corrosion was only produced during the wet period of 60 days the averaged corrosion rate acting was of about 11.3 $\mu A/cm^2$. This value of $I_{corr}$ calculated would have resulted smaller, if part of the corrosion was produced later during the 20 years period in which the beam

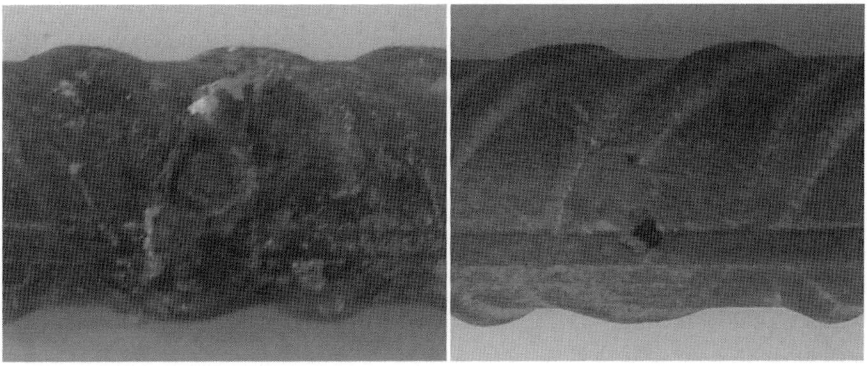

**Fig. 10** Aspect of actual local attack

was sheltered from rain, that is, if periods of water condensation could have been produced in spite of the sheltered conditions.

Just for the sake of analyzing the order of magnitude of the attack, if the corrosion would had been constant during 20 years, the averaged active corrosion rate would be of around 0.09 $\mu A/cm^2$ as calculated from the following Eq. 1:

$$P_{corr}(\mu m) = 0.0116 \cdot I_{corr} \cdot t \cdot \alpha = 215\mu m. \quad (1)$$

## 5 Conclusions

The work has consisted in analyzing the results of beams made around 20 years ago, to submit them to new aggressive exposure and to observe the performance.

- From the results after 20 years, it can be drawn up that in carbonated "dry" concretes, some corrosion spots were noticed in cracks greater than 0.08 mm.
- In concrete exposed to chloride solution during some months after their preparation (20 years ago), active corrosion was noticed in cracks greater than 0.4 mm. However, the corrosion rates were very low when measured after 20 years in outside sheltered conditions.

From the tests recently performed on the 20 years aged beams:

- In carbonated concrete with cracks greater than 0.2 mm, (no smaller cracks were studied), active corrosion in the crack region is noticed either when in contact with tap water or with sodium chloride solution.
- Regarding the localization of the attack, the corrosion penetration depth has been measured after these 20 years in the crack region of one of the beams only submitted initially to wet-dry cycles during 60 days (Table 2-stage 1). As the penetration depth measured was around 215 μm, the corrosion rates able to produce that attack depth in 60 days resulted of around 11.3 $\mu A/cm^2$.

**Acknowledgements** The authors are grateful to the grant funded by the Ministry of Economy and Competitivity of Spain through the research project INPACTO- EXTREPHOR

## References

Arya C, Ofori-Darko FK (1996) Influence of crack frequency on reinforcement corrosion in concrete. Cem Concr Res 26:345–353

Beeby AW (1978) Corrosion of reinforcing steel in concrete and its relation to cracking. Struct Eng 56:77–81

Beeby AW (1983) Cracking, cover and corrosion of reinforcement. Concr Int 5:35–40

Nürnberger U (1984) Chloride corrosion of steel in concrete, part 2. Betonwerk Fertigteil-technik 10:697–704

Otieno MB, Alexander MG, Beushausen HD (2010) Corrosion in cracked and uncracked concrete-influence of crack width, concrete quality and crack reopening. Mag Concr Res 62(6):393–404

Sarría Odiaga JJ (1997) Effect of the environmental parameters on the rate of corrosion of steel embedded in the concrete. Incidence of cover cracking doctoral thesis, Polytechnic University of Madrid

Schiessl P (1988) Cracking of concrete and durability of concrete structures. In: European Conference on cracking of concrete and durability of construction, Paris

Tarek MU et al (2001) Effect of crack width and bar types on corrosion of steel in concrete. J Mater Civil Eng 13(3):194–201

# Influence of Cracking Caused by Structural Loading on Chloride-Induced Corrosion Process in Reinforced Concrete Elements: A Review

**Junjie Wang, P.A.M. Basheer, S.V. Nanukuttan, and Yun Bai**

**Abstract** The chloride-induced corrosion of steel is a serious problem affecting the durability of reinforced concrete structures. Cracks may facilitate the ingress of chlorides into concrete and thus accelerate the corrosion process. However, different opinions on the effect of cracks on corrosion of steel were found in the literature. Two aspects existed among researchers: one is whether cracks accelerate only the corrosion initiation process or the overall corrosion process. The other aspect is whether or not there is a relationship between the crack width and the corrosion rate. In order to answer these contradictory aspects, finds from the literature were summarised. All the factors that could influence the corrosion process of steel in cracked concrete, such as crack inducing methods, corrosion acceleration methods, crack width, crack orientation, crack propagation status, concrete cover depth and concrete composition (water-cementitious materials ratio and mineral additives), were analysed based on the findings from the literature. A summary of current knowledge and conclusions about the aspects was then presented.

**Keywords** Crack • Chloride • Corrosion • Concrete • Durability

## 1 Introduction

Chloride-induced corrosion is one of the most important causes of the deterioration of reinforced concrete structures in marine environments and where deicing salt is used to melt ice in winter, because of ingress of chloride ions into reinforced

J. Wang (✉) • P.A.M. Basheer • S.V. Nanukuttan
School of Planning, Architecture and Civil Engineering, Queens University of Belfast, Belfast, United Kingdom
e-mail: Jwang06@qub.ac.uk

Y. Bai
Department of Civil, Environmental and Geomatic Engineering, University College London, London, UK

 
C. Andrade et al. (eds.), *Durability of Reinforced Concrete from Composition to Protection*, DOI 10.1007/978-3-319-09921-7_10

concrete. Although numerous investigations have been conducted in this area (Yoon et al. 2000), only few considered the influence of cracking on the chloride-induced corrosion process. In fact, cracking can easily occur when part of concrete is subjected to tensile forces; a typical example is a concrete beam under lateral loading. However, there are different opinions on the relationship between cracks and chloride-induced corrosion process of steel in reinforced concrete. Two different aspects not fully clarified about the influence of cracking on corrosion in reinforced concrete, as shown below, have existed for a long time (Otieno et al. 2010; Jaffer and Hansson 2008; Mohammed et al. 2001).

Aspect 1: Presence of cracks versus corrosion process

Opinion 1 – cracking accelerates both the initiation and propagation stages
Opinion 2 – cracking only accelerates the initiation stage but not at the later stage

Aspect 2: Crack width vs. corrosion rate

Opinion 1 – the wider the crack width the greater the corrosion rate of steel in cracked concrete
Opinion 2 – no relationship between crack width and corrosion rate of steel in cracked concrete

The first aspect is whether or not the presence of cracks accelerates the whole corrosion process or just the initiation stage (Otieno et al. 2010; Jaffer and Hansson 2008; Bremner et al. 2001; Arya and Wood 1995; Darwin et al. 1985). According to one set of findings (Zhang et al. 2010; Vidal et al. 2007), the effect of cracks on the overall corrosion process is not important, but only for the corrosion initiation process. The other opinion is that the overall process can be seriously affected (Sangoju et al. 2011; Otieno et al. 2010; Jaffer and Hansson 2008; Poursaee and Hansson 2008; Sahmaran and Yaman 2008; Saraswathy and Song 2007; Scott and Alexander 2007). The other aspect is whether or not there is any relationship between the crack width and the corrosion rate of steel in concrete (Mohammed et al. 2001). Some researchers (Sangoju et al. 2011; Otieno et al. 2010; Jaffer and Hansson 2008; Poursaee and Hansson 2008; Sahmaran and Yaman 2008; Saraswathy and Song 2007; Scott and Alexander 2007) found that a wider crack width resulted in greater corrosion rate, while others (Zhang et al. 2010; Vidal et al. 2007; Beeby 1983) got little relationship between the crack width and the corrosion rate.

In order to find the reasons for these differences, all the factors influencing corrosion of steel in cracked concrete, like crack inducing methods, corrosion acceleration methods, crack width, crack orientation, crack propagation status, concrete cover depth and concrete composition (water-cementitious ratio and mineral additives), were analysed based on published literature. Major conclusions were identified, and then, a summary of current knowledge in relation to the two aspects was given.

# 2 Factors Influencing Corrosion of Steel in Cracked Concrete

## 2.1 *Methods of Inducing Structural Cracks*

Different methods used can induce different crack patterns and the corrosion behaviour of steel in concrete can be different. As shown in Fig. 1, three-point and four-point bending are the most commonly used methods to induce transverse cracks on the tensile surface of the reinforced concrete beam. Other bending methods, like cantilever method (Saraswathy and Song 2007; Li 2001) as shown in Fig. 2 and U-shape bending method (Sangoju et al. 2011) as shown in Fig. 3, have the same principles as three-point or four-point bending method. The direction of a transverse crack is usually perpendicular to the longitudinal reinforcement. Longitudinal crack can be caused by the difference between thermal expansion coefficients of polymethylmethacrylate plastic bar and concrete as shown in Fig. 4 (Poursaee and Hansson 2008). Dynamic loading system (Jaffer and Hansson 2008) as shown in Fig. 5 can induce dynamic cracks. Artificial cracking method in Fig. 6 by inserting plates during casting was also used by some researchers (Montes et al. 2004; Arya and Ofori-Darko 1996).

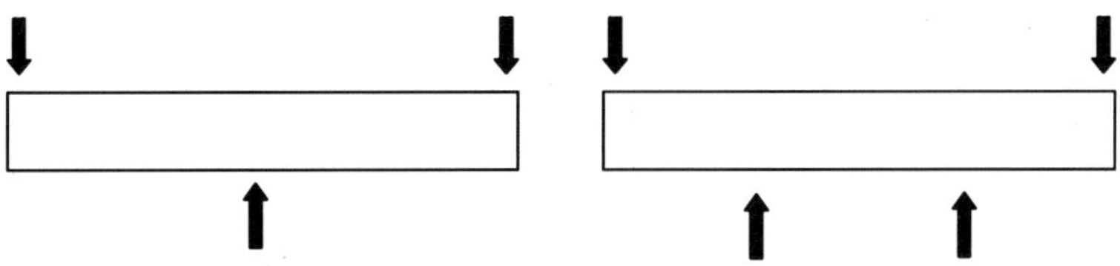

**Fig. 1** Three-point bending (*left*) and four-point bending (*right*) methods

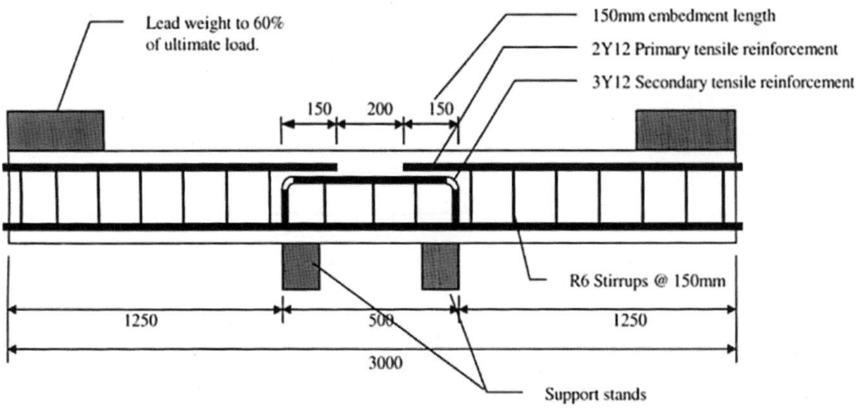


**Fig. 2** Cantilever bending method with sustained loading system (Li 2001)

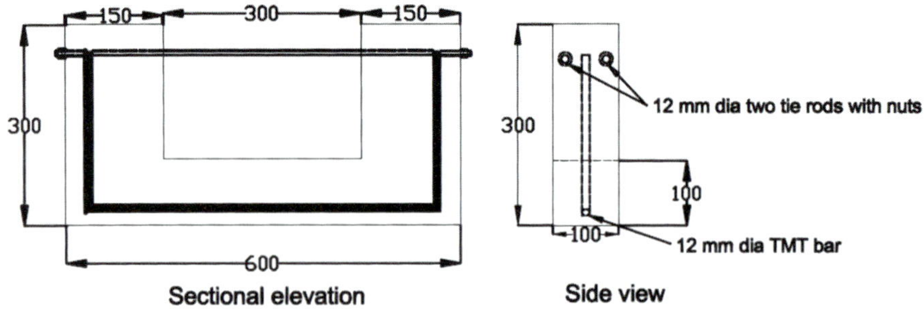


**Fig. 3** U shape bending method (Sangoju et al. 2011)

**Fig. 4** Longitudinal crack caused by thermal expansion coefficient difference between polymethylmethacrylate (*PMMA*) rod and concrete (Poursaee and Hansson 2008)

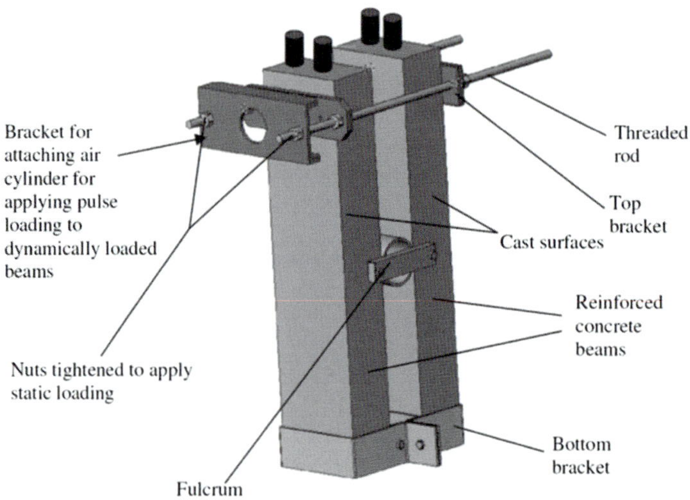


**Fig. 5** Dynamic loading system under three-point bending (Jaffer and Hansson 2008)

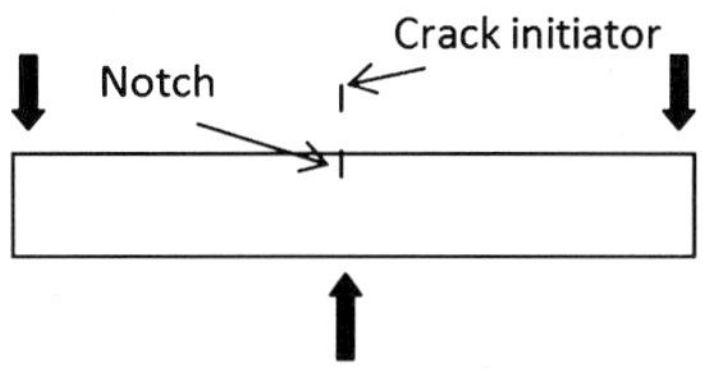


**Fig. 6** Single crack caused under three-point bending

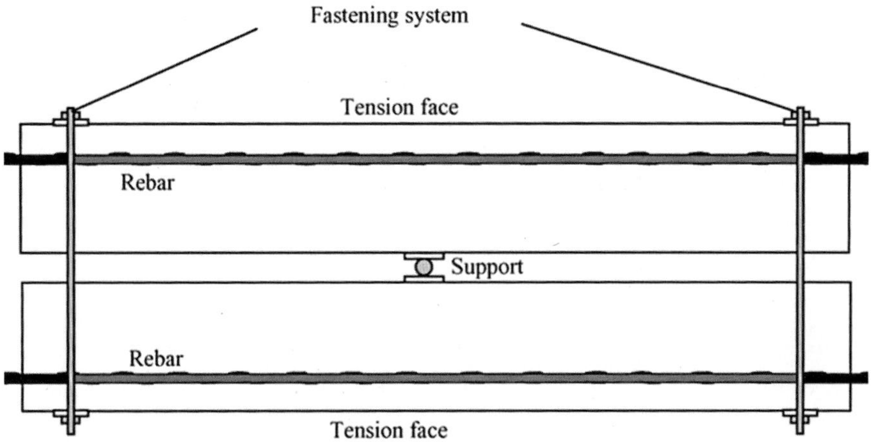


**Fig. 7** Fastening system under three-point bending (Cheng et al. 2005)

Experiments for the influence of cracks on corrosion of steel can be designed by either controlling the crack width or loading level. However, only initial crack width can be controlled, because the crack width can change with time due to the later corrosion process, creep and self-healing process. Controlling loading level is thus a better method. Loading level controlled by steel bolts, like Figs. 3 and 7, is just the initial loading level because of creep and stress relaxation in the steel bolt with time. Constant loading level can be achieved by putting actual weight on the free end of cantilever (Li 2001) as shown in Fig. 2 or on a lever arm (Yoon et al. 2000) as shown in Fig. 8.

## 2.2 *Methods of Corrosion Acceleration*

Different corrosion behaviours of steel in concrete can be found in tests conducted under different corrosion methods. Methods like wetting/drying with chloride solution, ponding and applying external electric field were used to accelerate corrosion process. No effect of cracks on corrosion of steel was found at any time in Pettersson's work (Pettersson 1996), the cracked reinforced concrete was submerged under the sea, this might be due to the limited availability of oxygen under the sea. Obvious differences on corrosion behaviour were reported between the

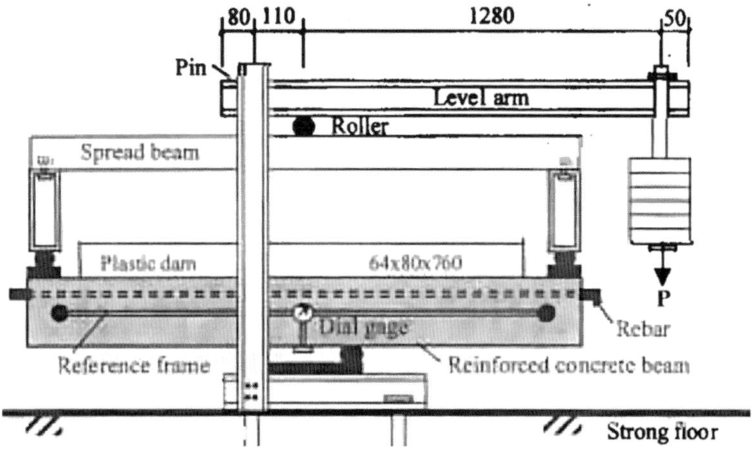


**Fig. 8** Four-point bending with sustained loading system (Yoon et al. 2000)

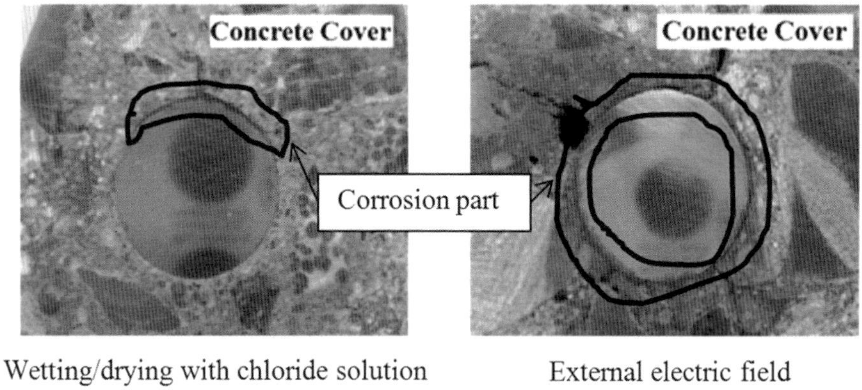


**Fig. 9** Different corrosion behaviours of steel in concrete under wetting/drying with chloride solution and under external electric field (Yuan et al. 2007)

tests conducted under external electric field and wetting/drying with chloride solution (Yuan et al 2007). As shown in Fig. 9, corrosion mainly occurred on the front part of steel surface facing chloride ingress in concrete under wetting/drying with chloride solution; however, corrosion on whole surface of steel was found for concrete under external electric field (Yuan et al 2007).

## 2.3 Crack Width

Table 1 shows the summary of findings from the literature on the effect of crack width on corrosion process of steel in concrete. It is interesting to see that, in the

**Table 1** Summary of literature on the effect of crack width on corrosion process

| Crack width | Test period | Aspect 1 | | Aspect 2 | |
|---|---|---|---|---|---|
| | | Opinion 1 | Opinion 2 | Opinion 1 | Opinion 2 |
| Small (≤0.5 mm) | Short (≤2 years) | • Sangoju et al. (2011), Otieno et al. (2010), Jaffer and Hansson (2008), Poursaee and Hansson (2008), Sahmaran and Yaman (2008), Saraswathy and Song (2007), Scott and Alexander (2007), Cheng et al. (2005), Montes et al. (2004), Mohammed et al. (2001, 2003), Yoon et al. (2000), Arya and Ofori-Darko (1996), Patil et al. (1991), Suzuki et al. (1990) | | • Sangoju et al. (2011), Otieno et al. (2010), Jaffer and Hansson (2008), Poursaee and Hansson (2008), Sahmaran and Yaman (2008), Saraswathy and Song (2007), Scott and Alexander (2007), Cheng et al. (2005), Montes et al. (2004), Mohammed et al. (2001, 2003), Yoon et al. (2000), Schiessl and Raupach (1997), Arya and Ofori-Darko (1996), Patil et al. (1991), Suzuki et al. (1990) | |
| | Long (≥12 years) | | • Zhang et al. (2010), Vidal et al. (2007), Francois and Arliguie (1991, 1998, 1999) | | • Zhang et al. (2010), Vidal et al. (2007), Francois and Arliguie (1991, 1998, 1999) |
| Large (≥1.0 mm) | Short (≤2 years) | • Pease et al. (2011), Arya and Ofori-Darko (1996), Patil et al. (1991) | | | |
| | Long (≥12 years) | Mohammed et al. (2003) | | Mohammed et al. (2003) | |

**Table 2** Crack width and maximum pitting corrosion depth after 15 years of exposure (Mohammed et al. 2003)

| Cement type | Initial crack width (mm), healing condition | Maximum pit depth at crack location (mm) | Maximum pit depth at uncracked regions (mm) |
|---|---|---|---|
| Ordinary Portland cement | 0.2, healed | 0.5 | 1.0 |
| | 0.1, healed | 0[a] | 0.5 |
| | 0.1, healed | 0.5 | 1.5 |
| Slag cement A | 0.1, healed | 0[a] | 0[a] |
| | 0.3, healed | 0.5 | 1.0 |
| | 0.2, healed | 0[a] | 0[a] |
| Slag cement B | 0.3, healed | 0.5 | 0[a] |
| | 0.1, healed | 0[a] | 0[a] |
| | 5.0, unhealed | 3.5 | 0[a] |
| Slag cement C | 1.5, unhealed | 1.0 | 0[a] |
| | 0.1, healed | 0[a] | 0[a] |
| | 2.0, unhealed | 0.5 | 0[a] |
| Fly ash cement B | 0.1, healed | 0.5 | 0.5 |
| | 0.3, healed | 1.0 | 0.5 |
| | 0.5, healed | 1.5 | 0.5 |

[a]Note pit depths <0.5 mm were not counted

case of reinforced concrete with small crack width (≤0.5 mm), all findings support Opinion 1 in Aspect 1 and Opinion 1 in Aspect 2 for the short term test (≤2 years), but for the long term test (≥12 years), all findings support Opinion 2 in Aspect 1 and Opinion 2 in Aspect 2. However, in the case of reinforced concrete with large crack width (≥1.0 mm), all results support Opinion 1 in Aspect 1 and Opinion 1 in Aspect 2 regardless of test duration. That is to say that, in reinforced concrete with small crack width (≤0.5 mm), the presence of cracks accelerated both the initiation and the propagation process and wider the crack width higher the corrosion rate of steel in the initial stage (≤2 years), but after long term exposure (≥12 years), no relationship between cracks or crack width and corrosion could be found. However, in reinforced concretes with large crack width (≥1.0 mm), the presence of cracks accelerated both the initiation and the propagation processes and wider the crack width higher the corrosion rate of steel both in the initial stage and in the long term exposure. Table 2 shows the results of work of Mohammed et al (2003), after exposure of reinforced concrete beams for 15 years in tidal zone, no effect of crack width below 0.5 mm on corrosion was found, but serious pitting corrosion on steel happened in cracks with crack width above 1.0 mm, while no corrosion happened in other locations in the same beam.

## 2.4 Crack Orientation

Crack orientations can be divided into two groups: transverse (cracks crossing main steel at right angles) and longitudinal (cracks parallel to the main steel, usually at

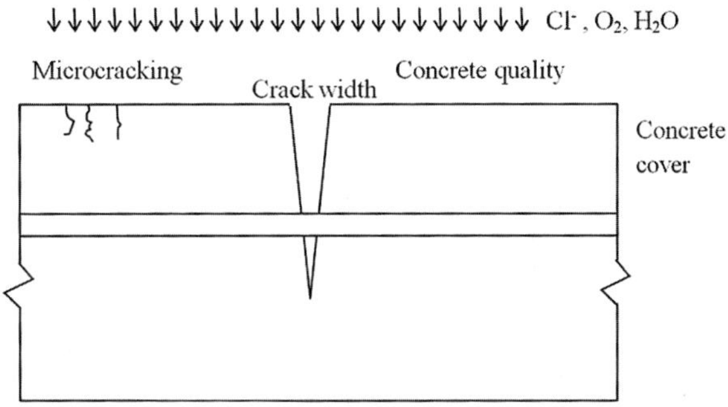


**Fig. 10** Diagram of reinforced concrete beam with transverse cracks

**Table 3** Summary of literature on the effect of crack orientation on corrosion process

| Crack orientation | Aspect 1 | |
|---|---|---|
| | Opinion 1 | Opinion 2 |
| Transverse | Sangoju et al. (2011), Otieno et al. (2010), Jaffer and Hansson (2008), Sahmaran and Yaman (2008), Saraswathy and Song (2007), Scott and Alexander (2007), Cheng et al. (2005), Montes et al. (2004), Mohammed et al. (2003, 2001), Yoon et al. (2000), Arya and Ofori-Darko (1996), Patil et al. (1991), Suzuki et al. (1990) | • Zhang et al. (2010), Vidal et al. (2007), Francois and Arliguie (1998, 1999, 1991) |
| Longitudinal | Poursaee and Hansson (2008) | |

the top the steel). As shown in Fig. 10, transverse cracks, which is more common for reinforced concrete elements subjected to lateral loading, can usually be found on the tension surface of reinforced concrete beam when it is under bending. Longitudinal cracks can be formed by the settlement of reinforcement when and after casting. As shown in Table 3, different opinions for Aspect 1 were found for reinforced concrete with transverse cracks. However, in the case of longitudinal cracks, only Opinion 1 for Aspect 1 was found. It is believed that longitudinal cracks can greatly accelerate the whole corrosion process of reinforcement and significantly reduce the service life of reinforced concrete structures (Arya and Wood 1995). It was found in Poursaee and Hansson's work (2008) that there was much more corrosion of steel under longitudinal cracks than that under transverse cracks and there was no difference of corrosion of steel under longitudinal crack between high performance concrete and ordinary concrete.

## 2.5 *Crack Propagation Status*

Most research focused on specimens under static loading, but variable loads are very common for reinforced concrete structures in practice. The opening and closing of

**Table 4** Crack type and propagation status (Arya and Wood 1995)

| Type of crack | Static | Dynamic |
|---|---|---|
| Plastic shrinkage | Yes | |
| Hydration temperature difference | Yes | |
| Shrinkage | Yes | Yes |
| Thermal expansion | | Yes |
| Pathological expansion (frost, alkali aggregate reaction) | | Yes |
| Settlement of foundations | Yes | Yes |
| Load | Yes | Yes |

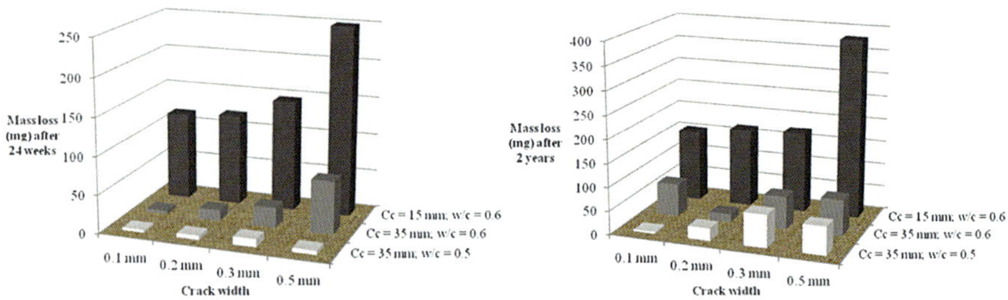


**Fig. 11** Calculated losses in mass of steel in crack zone due to macrocell corrosion after test period of 24 weeks and 2 years (Schiessl and Raupach 1997) (*cc* cover depth, *w/c* water-cementitious materials ratio)

cracks can influence chloride ingress and the damage state of reinforced concrete elements. Table 4 shows the propagation status of different types of cracks (Arya and Wood 1995). Different opinions about the effect of crack propagation status on corrosion were found in the literature. Otieno et al. (2010) reported that reloading of the cracked reinforced concrete increased the corrosion rate. However, Jaffer and Hansson's work (2008) suggested that the effect of type of loading on corrosion process was not that significant and less than that of type of concrete and exposure conditions.

## 2.6 Concrete Cover Depth

Schießl and Raupach (1997) observed much-reduced corrosion of steel by increasing the cover depth from 15 mm to 35 mm, as shown in Fig. 11. They concluded that concrete cover depth has a much greater influence than crack width on the corrosion in cracked reinforced concrete. Scott and Alexander (2007) also found that all corrosion rates were reduced by increasing the cover depth from 20 mm to 40 mm; however, this effect was only observed in ordinary Portland cement concrete (NC) but not in high performance concrete (HPC) with adding ground granulated blast-furnace slag (GGBS), pulverised fuel ash (PFA) and silica fume (SF). It is important to mention that by increasing the cover depth, the number of cracks could be reduced (Francois and Arliguie 1998; Arya and Ofori-Darko 1996) and the total corrosion might be reduced even more.

In Suzuki et al.'s work (1990), corrosion in cracked reinforced concrete specimens with cover depths of 20, 30 and 60 mm was investigated. They found that the effect of water-cementitious materials ratio (0.35–0.55) on corrosion in cracked reinforced concrete was greater than that of the cover depth.

## 2.7 Concrete Composition (Water-Cementitious Materials Ratio, Mineral Additives)

It was found that the corrosion rate in both cracked and uncracked reinforced concrete decreased with decreasing water-cementitious materials ratio (Otieno et al. 2008; 2010; Saraswathy and Song 2007; Montes et al. 2004; Mohammed et al. 2003; Mohammed et al. 2001; Suzuki et al. 1990). Some research indicated that the effect of water-cementitious materials ratio on corrosion is more important than that of crack width (Mohammed et al. 2001; Suzuki et al. 1990). Figure 12 (Otieno et al. 2010) shows a great increase of the corrosion rate, by increasing the water-cementitious ratio from 0.40 to 0.55, in ordinary Portland cement concretes with crack width from uncracked, incipient, 0.4 to 0.7 mm, and in and slag concretes with crack width 0.4 and 0.7 mm.

Many authors (Otieno et al. 2010; Jaffer and Hansson 2008; Scott and Alexander 2007; Montes et al. 2004; Sangoju et al. 2011; Saraswathy and Song 2007; Poursaee and Hansson 2008) have found that by adding mineral additives such as GGBS, PFA and SF, the corrosion rate of steel in cracked concrete can be significantly reduced. Gravimetric weight loss of rebars in ordinary Portland cement concrete

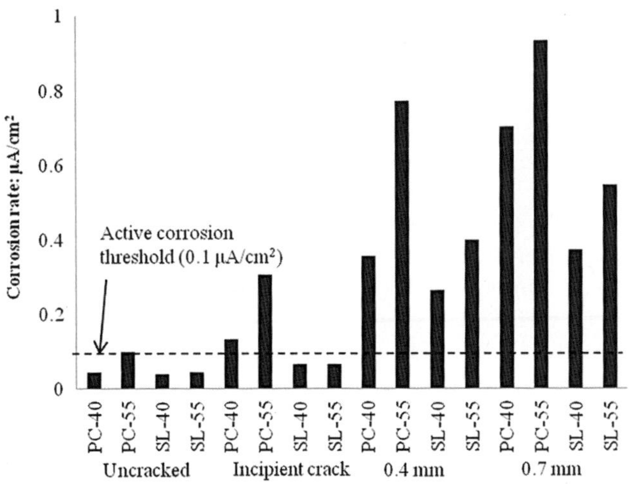


**Fig. 12** Average corrosion rates (weeks 26–31 under wetting/drying) (Otieno et al. 2010) (*PC*-ordinary Portland cement, *SL*-slag, 40 and 55 are the water-cementitious ratios of 0.40 and 0.55)

**Table 5** Summary of literature on the effect of concrete composition on corrosion process

| The better the concrete the lower the corrosion rate | Literature |
|---|---|
| Yes | Sangoju et al. (2011), Otieno et al. (2010), Jaffer and Hansson (2008), Saraswathy and Song (2007), Scott and Alexander (2007), Montes et al. (2004), Mohammed et al. (2003, 2001) |
| No | • Poursaee and Hansson (2008) – Longitudinal cracks |
| | • Mohammed et al. (2003) – Large crack widths ($\geq$1 mm) |

(NC) was about 1.5–2.0 times of that in PFA concrete (Sangoju et al. 2011). Scott and Alexander (2007) reported a 50 % reduction of corrosion rate in high performance concretes (HPCs) by adding GGBS, PFA and SF compared with NCs. This effect could be due to enhanced resistance to the transport of aggressive substances (chloride, water and oxygen), better crack healing and different crack path in the presence of these mineral additives (Poursaee and Hansson 2008).

However, as shown in Table 5, there are exceptions: concretes with large crack widths (Mohammed et al. 2003) or concretes with longitudinal cracks (Poursaee and Hansson 2008) do not support the relationship between quality of concrete and corrosion rate. Poursaee and Hansson (2008) performed corrosion studies about the influence of longitudinal cracks and transverse cracks on corrosion protection of NCs and HPCs. They observed that the HPCs provided a better protection of corrosion to steel under transverse cracks than NCs. However, no significant difference of corrosion was found between NCs and HPCs in the case of longitudinal cracks.

# 3 Conclusions

From the above analysis, the following conclusions have been drawn:

1. In the case of longitudinal cracks, aggressive substances can ingress directly into the concrete and reach the surface of the steel below the cracks, and the corrosion process can be greatly accelerated regardless of concrete composition and concrete cover depth.
2. In the case of transverse cracks, the corrosion process of steel in concrete is influenced by many factors, such as crack width, concrete cover depth and concrete quality. On the basis of information available in the literature, three models of the corrosion process can be summarised, as given in Table 6. Some old structures and poorly designed and cured concrete might be in Model 1 with poor quality concrete. In this case, the chloride ingress is very quick due to the poor quality of concrete and thus the overall corrosion process could also be very quick. The effect of cracks and crack width on corrosion of steel can be found only at the initial stage.

**Table 6** Three models of corrosion processes for steel under transverse cracks

| Model | Concrete quality | Crack width | Corrosion process |
|---|---|---|---|
| 1 | Poor[a] | Regardless | Cracking only influences corrosion at the initial stage. |
| | | | No effect of cracks after long term exposure |
| | | | Corrosion process can be very quick |
| 2 | Good/ excellent | Small | Cracking only influences corrosion at the initial stage |
| | | | No effect of cracks after long term exposure |
| | | | Corrosion process is determined by how long it takes for chloride to reach the steel level in uncracked parts, i.e. concrete cover depth |
| 3 | Good/ excellent | Large | Cracking accelerates the whole corrosion process |
| | | | Serious pitting corrosion and steel section loss can happen at the location of cracks |
| | | | Most significant loss of steel section can usually be found at the widest crack |

[a]Note that good quality of concrete with lots of microcracking still makes a poor concrete

Model 2, with good/excellent concrete quality and small crack width, might be the most common case for reinforced concretes exposed to aggressive environments. The influence of concrete cover depth and microcracking in uncracked parts on corrosion process of steel is more important than the presence of cracks and crack width. In this case, no effect of cracks on corrosion of steel can be found in long term exposure. Some improperly designed or overloaded structural elements might fit in Model 3, with the concrete the same as in Model 2 but with large crack width. This case can be the most dangerous situation for the reinforced concrete structures due to the sustained and very local attack of the corrosion of steel. The effect of cracks and crack width can be found throughout the deterioration process, with the most serious pitting corrosion found below the widest crack.

3. It can be summarised from above that, in order to have a long target service life of reinforced concrete structures, longitudinal cracks and large crack widths should be eliminated, and a good quality of concrete with a large cover depth and small crack width is needed.
4. The current knowledge for solutions of the aspects can be summarised as follows:

Aspect 1: Presence of cracks versus corrosion process

Opinion 1 – cracking accelerates both the initiation and propagation stages
Opinion 2 – cracking only accelerates the initiation stage but not the later stage

Solution:

Models 1 and 2 in Table 6 agree with Opinion 2
Model 3 in Table 6 and case of steel under longitudinal crack agree with Opinion 1

Aspect 2: Crack width versus corrosion rate

Opinion 1 – the wider the crack width the greater the corrosion rate of steel in cracked concrete

Opinion 2 – no relationship between crack width and corrosion rate of steel in cracked concrete

Solution:

Models 1 and 2 in Table 6 agree with Opinion 2 for the long term exposure. Opinion 1 can be found in initial stage

Model 3 in Table 6 agrees with Opinion 1

## References


Arya C, Ofori-Darko FK (1996) Influence of crack frequency on reinforcement corrosion in concrete. Cem Concr Res 26(3):345–353

Arya C, Wood LA (1995) Relevance of cracking in concrete to corrosion in concrete, Concrete Society Technical Report No 44

Beeby A (1983) Cracking, cover and corrosion of reinforcement. Concr Int 5(2):35–40

Bremner T, Hover K, Poston R, Broomfield J, Joseph T, Price R, Clear K, Khan M, Reddy D, Clifton J (2001) ACI 222R-01 protection of metals in concrete against corrosion. American Concrete Institute, Farmington Hills

Cheng A, Huang R, Wu J, Chen C (2005) Influence of GGBS on durability and corrosion behavior of reinforced concrete. Mater Chem Phys 93(2–3):404–411

Darwin D, Manning D, Hognestad E, Beeby A (1985) Aspect: crack width, cover, and corrosion. Concr Int 7(5):20–35

Francois R, Arliguie G (1991) Reinforced concrete: correlation between cracking and corrosion. Spec Publ 126:1221–1238

Francois R, Arliguie G (1998) Influence of service cracking on reinforcement steel corrosion. J Mater Civil Eng 10:14

Francois R, Arliguie G (1999) Effect of microcracking and cracking on the development of corrosion in reinforced concrete members. Mag Concr Res 51(2):143–150

Jaffer S, Hansson C (2008) The influence of cracks on chloride-induced corrosion of steel in ordinary Portland cement and high performance concretes subjected to different loading conditions. Corros Sci 50(12):3343–3355

Li C (2001) Initiation of chloride-induced reinforcement corrosion in concrete structural members – experimentation. ACI Struct J 98(4):502–510

Mohammed TU, Otsuki N, Hisada M, Shibata T (2001) Effect of crack width and bar types on corrosion of steel in concrete. J Mater Civil Eng 13(3):194–201

Mohammed TU, Otsuki N, Hamada H (2003) Corrosion of steel bars in cracked concrete under marine environment. J Mater Civil Eng 15(5):460–469

Montes P, Bremmer T, Lister D (2004) Influence of calcium nitrite inhibitor and crack width on corrosion of steel in high performance concrete subjected to a simulated marine environment. Cem Concr Compos 26(3):243–253

Otieno M, Alexander M, Beushausen H (2008) Corrosion propagation in cracked and uncracked concrete. In: Concrete repair, rehabilitation and retrofitting: 2nd international conference on concrete repair, rehabilitation and retrofitting, ICCRRR-2, 24–26 November 2008, Cape Town, South Africa, pp. 339–344

Otieno M, Alexander M, Beushausen H (2010) Corrosion in cracked and uncracked concrete-influence of crack width, concrete quality and crack reopening. Mag Concr Res 62(6):393–404

Patil B, Ranganna G, Channel S, Wagh A, Sawant S, Gajendragad M (1991) Effects of stress corrosion crack on steel in embedded concrete under marine conditions. In: 4th Indian conference on ocean engineering, 4–6 September 1991, NIO, Goa, pp. 499–505

Pease B, Geiker M, Stang H, Weiss J (2011) The design of an instrumented rebar for assessment of corrosion in cracked reinforced concrete. Mater Struct 44(7):1259–1271

Pettersson K (1996) Criteria for cracks in connection with corrosion in high-strength concrete. In: de Larrard F, Lacroix R (eds) The 4th international symposium on utilization of high-strength/ high-performance concrete, Paris, p 509

Poursaee A, Hansson CM (2008) The influence of longitudinal cracks on the corrosion protection afforded reinforcing steel in high performance concrete. Cem Concr Res 38(8–9):1098–1105

Sahmaran M, Yaman IÖ (2008) Influence of transverse crack width on reinforcement corrosion initiation and propagation in mortar beams. Can J Civil Eng 35(3):236–245

Sangoju B, Gettu R, Bharatkumar BH, Neelamegam M (2011) Chloride-induced corrosion of steel in cracked OPC and PPC concretes: experimental study. J Mater Civil Eng 23(7):1057–1066

Saraswathy V, Song H (2007) Evaluation of corrosion resistance of Portland pozzolana cement and fly ash blended cements in pre-cracked reinforced concrete slabs under accelerated testing conditions. Mater Chem Phys 104(2–3):356–361

Schiessl P, Raupach M (1997) Laboratory studies and calculations on the influence of crack width on chloride-induced corrosion of steel in concrete. ACI Mater J 94(1):56–62

Scott A, Alexander MG (2007) The influence of binder type, cracking and cover on corrosion rates of steel in chloride-contaminated concrete. Mag Concr Res 59(7):495–505

Suzuki K, Ohno Y, Praparntanatorn S, Tamura H (1990) Mechanism of steel corrosion in cracked concrete. In: Page CL, Treadaway K, Bamforth PB (eds) Corrosion of reinforcement in concrete. In: 3rd international symposium on corrosion of reinforcement in concrete construction, Wishaw, Warwickshire, 21–24 May, 1990, pp 19–28

Vidal T, Castel A, François R (2007) Corrosion process and structural performance of a 17 year old reinforced concrete beam stored in chloride environment. Cem Concr Res 37(11):1551–1561

Yoon S, Wang K, Weiss WJ, Shah SP (2000) Interaction between loading, corrosion, and serviceability of reinforced concrete. ACI Mater J 97(6):637–644

Yuan Y, Ji Y, Shah SP (2007) Comparison of two accelerated corrosion techniques for concrete structures. ACI Struct J 104(3):344–347

Zhang R, Castel A, François R (2010) Concrete cover cracking with reinforcement corrosion of RC beam during chloride-induced corrosion process. Cem Concr Res 40(3):415–425

# Effect on Mechanical Properties and Chloride Penetration Resistance of Modified Hydrotalcite in Cement Mortar

**Zhengxian Yang, Hartmut Fischer, and Rob Polder**

**Abstract** In this paper, two types of modified hydrotalcites (MHT-pAB and MHT-$NO_2$) were incorporated into cement mortars with two dosage levels (replacing 5 and 10 % cement) and a constant water-to-(cement + MHT) ratio of 0.50. Designated testing programme including strength test, porosity test, and rapid chloride migration were employed to investigate the effect of modified hydrotalcites on chloride penetration in cement mortar. The results based on these tests revealed that the incorporation of MHT-pAB at 5 % dosage in mortar specimens produced a notably improved resistance to chloride ingress with no or minor influence on the development of mechanical strength, which further confirmed the possibilities for using modified hydrotalcite as a chloride scavenger in cement mortars.

**Keywords** Modified hydrotalcite • Reinforced concrete • Mechanical properties • Chloride penetration

## 1 Introduction

Exposed to service environment, the durability of concrete can be compromised by the ingress of water with dissolved corrosive ions, chlorides, and other deleterious species which cause corrosion of the reinforcing steel. This occurs via deterioration of the passivation layer of the steel reinforcement, presented only in the high

---

Z. Yang (✉)
Materials Innovation Institute (M2i), P.O. Box 5008, 2600 GA Delft, The Netherlands

Section of Materials and Environment, Faculty of Civil Engineering and Geosciences, Delft University of Technology, P.O. Box 5048, 2600 GA Delft, The Netherlands
e-mail: Zhengxian.Yang@tudelft.nl

H. Fischer
TNO Materials Performance, P.O. Box 6235, 5600 HE Eindhoven, The Netherlands

R. Polder
Section of Materials and Environment, Faculty of Civil Engineering and Geosciences, Delft University of Technology, P.O. Box 5048, 2600 GA Delft, The Netherlands

TNO Structural Reliability, P.O. Box 49, 2600 AA Delft, The Netherlands

 
C. Andrade et al. (eds.), *Durability of Reinforced Concrete from Composition to Protection*, DOI 10.1007/978-3-319-09921-7_11

alkaline environment of concrete. Among the aggressive species which could potentially impose a serious threat on the durability and serviceability of reinforced concrete structures (Bertolini et al. 2013), chloride is the most detrimental culprit and has been recognized as the one of the major factors affecting the service life of reinforced concrete structures (Polder and Peelen 2002; Polder and Hug 2000; Yang et al. 2009). According to Tuutti's corrosion model, the development of corrosion in reinforced concrete structures can be divided in two main stages (Tuutti 1982): the first stage is the initiation of corrosion in which the reinforcement is passive but phenomena that can lead to loss of passivity, e.g. chloride penetration into the concrete cover take place; the second stage is the corrosion propagation which is related to reinforcement corrosion and concrete cover cracking and spalling. Most traditionally available corrosion prevention approaches such as coatings on reinforcing steel, concrete surface sealer, stainless steel reinforcement and cathodic protection (Cigna et al. 2002; Elsener et al 2010; Elsener 2001; Pedeferri 1996) sometimes work efficiently only in one of the two stages but limit to the other one. Thus continuing research in the domain of materials science is essentially needed in searching for more effective measures to improve the corrosion resistance of reinforced concrete.

A possibility is to use modified hydrotalcite as an alternative approach for chloride scavenging in concrete (Yang et al. 2013a, 2014). Hydrotalcite is one representative of a large mineral group of Layered Double Hydroxides (LDHs), in general formula $[M^{II}_{1-x}M^{III}_{x}(OH)_2]^{x+}[(A^{n-}_{x/n})]^{x-}\bullet mH_2O$, where $M^{II}$ and $M^{III}$ are di- and trivalent metals respectively and $A^{n-}$ is an exchangeable interlayer anion with valence n. The x value is in the range of 0.22–0.33. A schematic of their structure and more detailed information related to MHTs' molecular structure and chemical composition can be found elsewhere (Yang et al. 2013b). Due to the unique characteristic of the molecular structure and their high anionic exchange capacity, hydrotalcite is believed to have a potential to be modified or tailor-made as an active component of concrete. Hydrotalcite have been found in hydrated slag cements, which are known to bind more chloride ions than pure Portland cements (Dhir et al.1996; Arya and Xu 1995; Glass and Buenfeld 2000). More recently, Kayali et al. (2012) demonstrated the remarkable role of hydrotalcite in chloride binding and corrosion protection in GGBFS concretes. A distinctive feature of MHTs relative to the other traditional corrosion prevention approaches is their dual function working mechanism: capturing aggressive chlorides and simultaneously releasing inhibitive anions to protect the reinforcing steel from corrosion (Yang et al. 2013b). Previous studies (Yang et al. 2013a, 2014) conducted in chloride-containing alkaline solution have shown that ion exchange indeed occurred between chlorides and intercalated inhibitors in MHT's structure. The simultaneously released inhibitors were found to exhibit the envisioned inhibiting effect and caused corrosion initiation of the steel to shift to higher chloride concentrations. Following those positive results obtained in chloride-containing alkaline solution, this paper aims to investigate the effect of modified hydrotalcites on chloride penetration in plain cement mortars.

# 2 Experimental

## 2.1 *Materials*

A European specification CEM I 42.5 N Portland cement in accordance with EN 197–1(2011), CEN-Standard sand (particle size: 0–2 mm) conforming to EN 196–1 (2005) and deionized water were used. MHTs that were used are synthetic Mg(2) Al-pAB (Mg/Al atomic ratio 2:1) and Mg(2)Al-$NO_2$. The two types of MHTs were synthesized by the modification of Mg(2)Al-$CO_3$ with sodium p-aminobenzoate (−pAB) and sodium nitrite (−$NO_2$) respectively through a calcination-rehydration procedure[11]. NaOH, $NaNO_2$ and p-aminobenzoic acid were obtained from Sigma-Aldrich. The hydrotalcite (Mg(2)Al-$CO_3$) used in this study is a commercially available product PURAL® MG 63 HT provided by Sasol Germany GmbH.

## 2.2 *Sample Preparation*

The mortar samples were prepared with a constant water-to-binder (cement + MHT, in this case MHT is counted as one of the binders) mass ratio of 0.50, a constant cement + MHT content, a constant sand-to-binder mass ratio of three and a MHT-to-cement mass ratio of 0 %, 5 %, 10 %, respectively. For both the reference (without MHT admixed) and mortar specimens admixed with MHTs, three specimens were prepared to ensure the statistical reliability of test results. For the specimens admixed with MHTs, the MHT powder was mixed first with dry cement in a mixer, and then sands were added and stirred thoroughly at a low speed for 30 s. Afterwards, deionized water was added into the mixture and stirred for about 90 s (low speed for 30 s and high speed for 60 s) to achieve good workability. After mixing, the fresh mixtures were cast in the specific molds and were carefully compacted on a vibrating table to minimize the amount of entrapped air. All the specimens were demolded after curing under room temperature (RT) and local lab environment for 24 h. Afterwards, they were moved to a fog room under $23 \pm 2$ °C and over 95 % relative humidity and continually cured for 2, 6 or 27 additional days before subjecting to any assigned testing.

# 3 Testing Methods

## 3.1 *Mechanical Test*

The mechanical properties especially the compressive and flexural strength were tested at curing ages of 3, 7 and 28 days by a standard three-point bending test using $40 \times 40 \times 160$ mm mortar prisms. The test was carried out in accordance with EN 196–1. The flexural strength was measured by applying the load vertically to the mortar prisms at the rate of (50 ± 10)N/s by a MACBEN(Servo plus evolution) materials testing equipment. The samples used for measuring the compressive strength were from the prism halves from the flexural strength test and the compression load was applied at the rate of (2,400 ± 200)N/s. At least three replicates were performed for both of the tests.

## 3.2 Mercury Intrusion Porosimetry

Mercury Intrusion Porosimetry (MIP) method is used to determine the porosity of mortar specimens admixed with/without MHTs by a "Micrometritics Poresizer 9500" mercury intrusion porosimeter. The porosity measurement was carried out in two stages. The first stage is at low pressure: from 0 to 0.0036 MPa. The second stage is the high pressure running from 0.0036 to 210 MPa and followed by an extrusion running from 210 to 0.14 MPa. The MIP samples were taken from the intact parts of the mortar samples after the mechanical test. The samples were first crushed by hammer into small pieces with dimensions of 1–2 $cm^3$. Then these pieces were immerged in liquid nitrogen for about 3 min to stop the hydration and stored in a vacuum freeze-dryer at −28 °C for more than two months until a stable mass loss of 0.01 %/day was reached before conducting the test. The quick freezing and drying process at low temperatures allows the remaining liquid solution transforming into ice microcrystals and further removed by sublimation without significant damage to the microstructure.

## 3.3 Rapid Chloride Migration (RCM) Test

The RCM tested was conducted following the procedures described in nordtest method NT Buid 492 (NT Build 492 1999) to determine the chloride migration coefficient of the mortar. After curing for 28 days, the 100 × 85 mm mortar cylinders were taken from the fog room. Two mortar slices of 15 and 20 mm were cut respectively from top and bottom of the specimen and consequently a 100 × 50 mm mortar slice was obtained as test specimen. The newly cut top surface is the one to be exposed to the chloride solution. The applied voltage is selected based on the initial current flowing through the test specimen at a voltage of 30 V. The higher is this initial current, the lower is the applied voltage. In this study, voltages of 20 and 15 V were selected respectively for the reference and mortar specimens admixed with MHTs based on the obtained initial current. Test duration was in these cases always 24 h.

# 4 Results and Discussion

## *4.1 The Effect of MHT on Strength Development*

Strength is an important engineering parameter to check the quality of concrete in its hardened state. According to EN-197, the standard compressive strength of a CEM I 42.5 N cement at 28 days is at least 42.5 MPa and hence this could be considered as a way to check the quality of cementitious mixtures made from this type of cement. In addition, Concrete with low porosity usually has high strength and high resistance to the penetration of aggressive ions, such as chlorides. Therefore, strength and diffusion coefficient data are sometimes linked together to the porosity of the concrete in order to globally assess the effect of a certain material on penetration of chloride. In this study, the effect of the MHTs on the development of compressive strength and flexural strength were studied. The results of the compressive strength and flexural strength test are shown respectively in Figs. 1 and 2, where each value is averaged from the results of three individual tests. As shown from these figures, comparing to the reference specimens (i.e., Ref.), the incorporation of MHTs results in observable but relatively small strength reduction at all the test ages. However, all of the compressive strength at 28 days is higher than 42.5 Mpa, a standard value that is recommended in EN-197. Therefore, one may conclude that no to minor negative effect on the strength development would result from the incorporation of MHTs in the cement mortar. Among the samples admixed with MHTs, the use of MHT-pAB at 5 % dosage performed the best in general. Additionally, the mortar samples admixed with MHT at 5 % dosage presented higher compressive and flexural strength values than those at 10 % dosages at the three test ages.

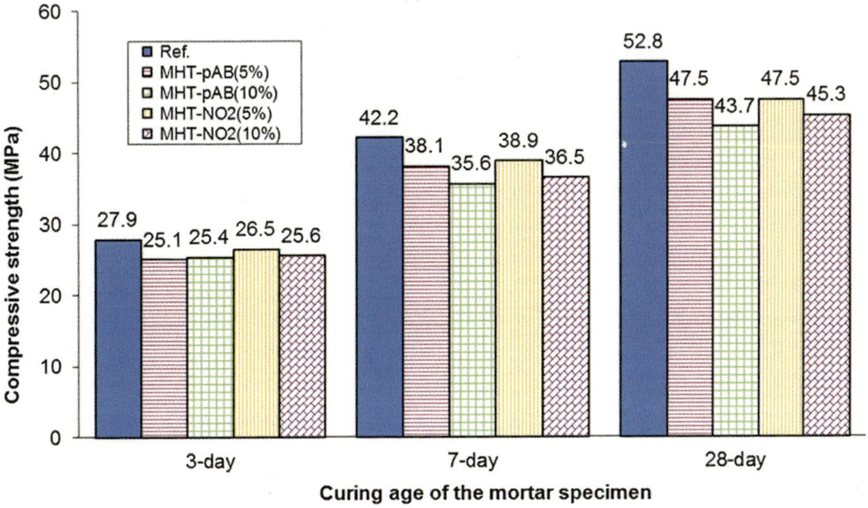


**Fig. 1** Compressive strength of the mortar specimens admixed with/without MHTs

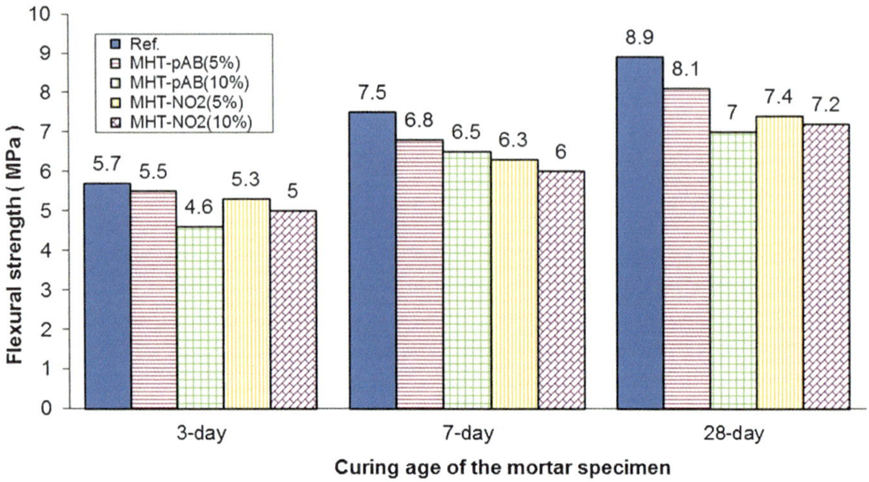


**Fig. 2** Flexural strength of the mortar specimens admixed with/without MHTs

## *4.2 The Effect of MHT on Porosity*

The MIP test has been performed on mortar samples at curing ages of 3, 7 and 28 days. The Washburn equation (Eq. 1) as shown below is used to calculate the diameter of pores intruded by mercury at each pressure step.

$$D = -\frac{4\gamma \cos\theta}{P} \tag{1}$$

where D is the pore diameter (m), is the surface tension of mercury (N/m), is the contact angle between mercury and the solid materials and P is the applied pressure (Pa). The MIP test procedure includes intrusion and extrusion. For cementitious material, the surface tension value of 480 (N/m) and the contact angle of $13\mathring{9}$ (Cook and Hover 1991) are suggested for the intrusion process and the contact angle of $10\mathring{6}$ (Léon and León 1998) is suggested for the extrusion process. According to the Washburn equation, the pore size ranging from 350 to 0.007 μm can be detected. Figure 3 shows the total porosity of all the test mortar specimens in which each value is the average of two parallel measurements. As can be seen from Fig. 3, relative to the Ref. specimen, the total porosity of the mortar has been increased in all the three test ages due to the incorporation of the MHTs in both of the two different dosage levels (5 and 10 %) with the exception of MHT-$NO_2$(10 %) whose porosity at 3 days, however, presented a relatively lower value. The increased porosity indicates the incorporation of the two types of MHTs has made the mortar a little more porous which in turn may increase the risk for chloride ion penetrating through the mortar matrix without considering the chloride binding behavior. As also seen from Figs. 1 and 2, the increased porosity has indeed led to some loss

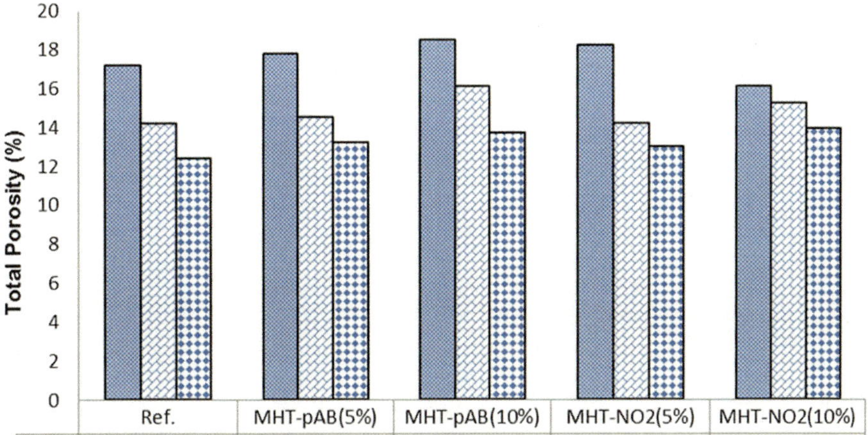


| | Ref. | MHT-pAB(5%) | MHT-pAB(10%) | MHT-NO2(5%) | MHT-NO2(10%) |
|---|---|---|---|---|---|
| 3d | 17.2 | 17.8 | 18.5 | 18.2 | 16.1 |
| 7d | 14.2 | 14.5 | 16.1 | 14.2 | 15.2 |
| 28d | 12.4 | 13.2 | 13.7 | 13.0 | 13.9 |

**Fig. 3** Porosity of the mortar specimens admixed with/without MHTs

of the strength although minor. In addition, when the results from the strength and porosity test are linked together, it is found that the development of the strength and porosity of the mortar specimens admixed with MHTs followed the same trend with the curing time (i.e., strength increased and porosity decreased as hydration proceed) as those of reference specimens which might indicate the incorporation of MHTs has no or very minor influence on the cement hydration process.

## *4.3 RCM Test*

Once the tests finished, all specimens were split axially into two pieces and sprayed 0.1 M silver nitrate ($AgNO_3$) solution immediately onto the freshly split mortar pieces to visualize the penetration depth of chloride. The average penetration depth of chloride was recorded accordingly. The non-steady-state migration coefficient ($D_{RCM}$) was then calculated by the following Eq. 2 proposed by NT Build 492.

$$D_{RCM} = \frac{0.0239(273+T)L}{(U-2)t}\left(x_d - 0.0238\sqrt{\frac{(273+T)Lx_d}{U-2}}\right) \quad (2)$$

Where $D_{RCM}$ is the non-steady-state migration coefficient, $\times 10^{-12}$ $m^2/s$; $X_d$ is the average of the chloride penetration depths, mm; T is average of the initial and final temperatures in anolyte solution, °C; U is the applied voltage V; L is the thickness of the specimen, mm; t is the test duration, h. The chloride migration coefficient

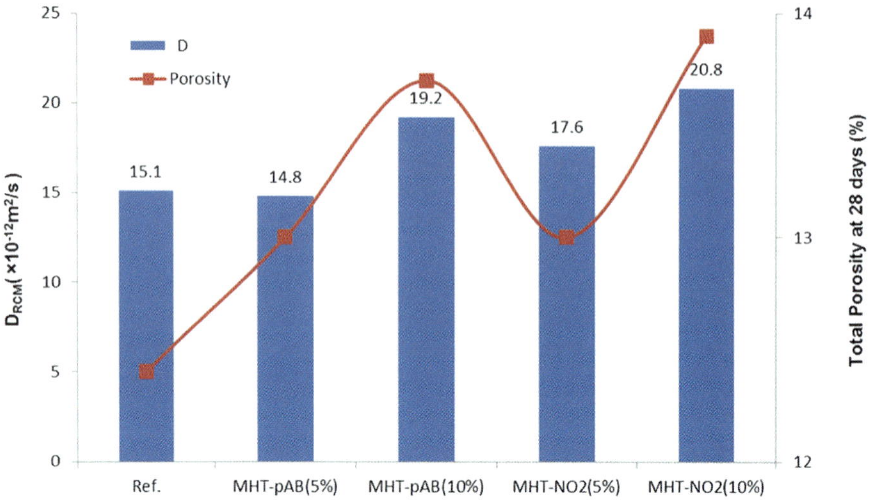


**Fig. 4** Chloride diffusion coefficient obtained from RCM test (bars and numbers) and porosity at 28 days obtained from MIP test (*squares* and *line*)

determined by RCM test has been widely used as one of the important parameters for resistance of cementitious materials to chloride penetration. Figure 4 shows the RCM test results averaged from three test specimens combined with the porosity results from MIP test at a curing age of 28 days.

As can be seen from Fig. 4, mortar specimen of MHT-pAB(5 %) presented the lowest RCM coefficient among all specimens that were tested and on the other hand, when compared with the Ref. specimens presented a relatively higher porosity. The co-occurrence of lower DRCM and higher porosity of mortar specimen admixed with MHT-pAB(5 %) may be attributed to an active chloride binding behavior that exists in its matrix. The active chloride binding behavior in the MHT-pAB(5 %) specimens could be traced back to the high anionic exchange capacity of MHTs which makes exchange of the interlayer ions by the intruded chlorides easily achieved as the following Eq. 3 shows(Yang et al. 2013c).

$$\text{MHT-}Inh + \text{Cl}^- \rightarrow \text{MHT-Cl} + Inh^- \tag{3}$$

Where -Inh represents the interlayer ions in the molecular structure of MHT (i.e., −pAB and $-NO_2$ anions in these cases). For the other MHT samples, the higher porosity may dominate the chloride migration process by the externally imposed electrical field although the ion exchange reactions might have occurred as well.

## 5 Conclusions

In this paper, two types of modified hydrotalcites (MHT-pAB and MHT-$NO_2$) were incorporated into cement mortars with two dosage levels (replacing cement at 5 and 10 % by mass). The mortar samples were prepared with a constant water-to-binder (cement + MHT) mass ratio of 0.50 and a constant sand-to-binder mass ratio of three. The mortar properties such as compressive and flexural strength, total porosity, and rapid chloride migration were investigated to shed light on a global picture of the effect of the two types of MHTs on chloride penetration in cement mortars and to study the potential correlation between these properties. Strength tests revealed that all of the compressive strength of the mortar specimens admixed with the two types of MHTs at 28 days is higher than 42.5 Mpa which is recommended by EN-197. Therefore, one may conclude that no to minor negative effect on the strength development would result from the incorporation of MHTs in the cement mortar. RCM test showed that mortar specimens admixed with MHT-pAB presented lower $D_{RCM}$ values at the two respective dosage levels than those mortar specimens admixed with MHT-$NO_2$ which may indicate that MHT-pAB has more active chloride binding behavior than MHT-$NO_2$ considering the relatively small differences between the porosity of MHT-pAB and MHT-$NO_2$ specimens at 28 days. In addition, the co-occurrence of lower chloride diffusion coefficient, $D_{RCM}$ and higher porosity of MHT-pAB(5 %) specimen comparing to the reference specimen suggested a higher chloride binding capacity of the mortar due to the incorporation of MHTs, which further indicated that ion exchange indeed occurred between chlorides and intercalated inhibitors (−pAB in this case) in MHT's structure through the dual function working mechanism of MHTs (Yang et al. 2013c). Nevertheless, this study validated the possibilities for using modified hydrotalcite as a chloride scavenger in cement mortars when an appropriate mixing dosage is adopted, which would subsequently results in slower chloride transport, thus delaying initiating the corrosion of the reinforcing steel.

**Acknowledgements** The research was carried out under the project number M81.609337 in the framework of the Research Program of the Materials innovation institute (M2i) (www.m2i.nl). Sasol Germany GmbH is acknowledged for the gift of hydrotalcite that was used in this research.

## References

Arya C, Xu Y (1995) Effect of cement type on chloride binding and corrosion of steel in concrete. Cem Concr Res 25(4):893–902

Bertolini L, Elsener B, Pedeferri P, Redaelli E, Polder R (2013) Corrosion of steel in concrete: prevention, diagnosis, repair. Wiley-VCH, Weinheim

Cigna R, Andrade C, Nürnberger U, Polder R, Weydert R, Seitz E (eds) (2002) COST 521: Final report, Luxembourg

Cook RA, Hover KC (1991) Experiments on the contact angle between mercury and hardened cement paste. Cem Concr Res 21(6):1165–1175

Dhir RK, El-Mohr MAK, Dyer TD (1996) Chloride binding in GGBS concrete. Cem Concr Res 26 (12):1767–1773

Elsener B (2001) Corrosion Inhibitors for steel in concrete-state of the art report, vol 35, EFC publication. IOM Communications, London

Elsener B, Addari D, Coray S, Rossi A (2010) Stainless steel reinforcing bars-reason for their high pitting corrosion resistance. Mater Corros 61:1–9

EN 196–1 (2005) Methods for testing cement-part 1: determination of strength. European standard
EN 197–1 (2011) Cement-part 1: composition, specification and conformity criteria for common cements. European standard
Glass GK, Buenfeld NR (2000) The influence of chloride binding on the chloride induced corrosion risk in reinforced concrete. Corros Sci 42(2):329–344
Kayali O, Khan MSH, Ahmed MS (2012) The role of hydrotalcite in chloride binding and corrosion protection in concretes with ground granulated blast furnace slag. Cem Concr Compos 34(8):936–945
Léon Y, León CA (1998) New perspectives in mercury porosimetry. Adv Colloid Interface Sci 76–77:341–372
NT Build 492 (1999) Concrete, mortar and cement-based repair materials: chloride migration coefficient from non-steady-state migration experiments. NordTest, Espoo
Pedeferri P (1996) Cathodic protection and cathodic prevention. Constr Bldg Mater 10(5):391–402
Polder R, Hug A (2000) Penetration of chloride from de-icing salt into concrete from a 30 year old bridge. Heron 45(2):109–124
Polder R, Peelen WHA (2002) Characterisation of chloride transport and reinforcement corrosion in concrete under cyclic wetting and drying by electrical resistivity. Cem Concr Compos 24 (5):427–435
Tuutti K (1982) Corrosion of steel in concrete. Swedish Foundation for Concrete Research (CBI), Stockholm
Yang Z, Shi X, Creighton AT, Peterson MM (2009) Effect of styrene-butadiene rubber latex on the chloride permeability and microstructure of Portland cement mortars. Constr Bldg Mater 23 (6):2283–2290
Yang Z, Fischer H, Polder R (2014) Synthesis and characterisation of modified hydrotalcites and their ion exchange characteristics in chloride-rich simulated concrete pore solution. Cem Concr Compos 47:87–93
Yang Z, Fischer H, Cerezo J, Mol JMC, Polder R (2013a) Aminobenzoate modified Mg-Al hydrotalcites as a novel smart additive of reinforced concrete for anticorrosion applications. Constr Bldg Mater 47:1436–1443
Yang Z, Fischer H, Polder R (2013b) Modified hydrotalcites as a new emerging class of smart additive of reinforced concrete for anti-corrosion applications: a literature review. Mater Corros 64(12):1066–1074

# Effects of Nano-silica (NS) Additions on Durability of SCC Mixtures

**G. Quercia, P. Spiesz, and H.J.H. Brouwers**

**Abstract** In this study, three different types of nano-silica were applied in self-compacting concrete (SCC), one produced by the controlled dissolution of the olivine mineral and two having similar particle size distributions (PSD), but produced through two different processes: fumed powder nano-silica and precipitated silica in colloidal suspension. The influence of the nano-silica on SCC was investigated with respect to the properties of the concrete in fresh (workability) and hardened state (durability properties). Additionally, the densification of the microstructure of the hardened concrete was analyzed by SEM and EDS techniques. The obtained results demonstrate that an efficient use of nano-silica in SCC can improve its durability properties. Considering the reactivity of the different nano-silica studied, the colloidal type showed a higher reactivity at early age, which influenced the final SCC properties.

**Keywords** Self-compacting concrete • Nano-silica • Olivine • Chloride • Freeze-thaw

## 1 Introduction

Currently, the micro-level does not provide enough insights into building materials. Therefore, all around the world, increasing amounts of funding are being directed to research projects dealing with material properties on nano-level, which is claimed to have tremendous potential for the future (Scrivener and Kirkpatrick 2008).

G. Quercia (✉)
Trican Well Service Ltd, Calgary R&D Centre, Canada
e-mail: gquercia@trican.ca

P. Spiesz
Department of the Built Environment, Eindhoven University of Technology, Eindhoven, The Netherlands

ENCI HeidelbergCement Benelux, The Netherlands
e-mail: p.spiesz@tue.nl; przemek.spiesz@heidelbergcement.com

H.J.H. Brouwers
Faculty of the Built Environment, Eindhoven University of Technology, Eindhoven, Netherlands

 
C. Andrade et al. (eds.), *Durability of Reinforced Concrete from Composition to Protection*, DOI 10.1007/978-3-319-09921-7_12

The fundamental processes that govern the properties of concrete are affected by the performance of the material on nano-scale. The main hydration product of cement-based materials, the CSH-gel, is a nano-structured material (Scrivener and Kirkpatrick 2008; Sanchez and Sobolev 2010). The mechanical properties and the durability of concrete mainly depend on the refinement of the microstructure of the hardened cement paste and the improvement of the paste-aggregate interface zone – ITZ – (Nili et al. 2010).

One of the most referred to and used cementitious nano-materials is amorphous silica with a particle size in the nano-range, even though its application and effects on concrete have not been fully understood yet. It has been reported that the nano-silica addition increases the compressive strength and reduces the overall permeability of hardened concrete due to its pozzolanic properties, which result in finer hydrated phases (CSH-gel) and densified (nano-filler and anti-$Ca(OH)_2$-leaching effect) microstructure (Rols et al. 1999; Ji 2005; Sobolev and Ferrara 2005; Lin et al. 2008; Gaitero et al. 2008; Belkowitz and Armentrout 2010; Quercia and Brouwers 2010). These effects may enhance the durability of concrete elements and structures.

There are different commercial types of nano-silica additives available on the market which are produced in different ways such as precipitation from olivine mineral dissolution, pyrolysis, sol–gel and others (Quercia and Brouwers 2010). The main characteristics of nano-silica, such as the particle size distribution, specific density, specific surface area, pore structure, and reactivity (surface silanol groups), depend on the production method. Despite the existence of several studies that describe the main properties and characteristics of concrete containing nano-silica particles, most of them focus on the application of nano-silica as an anti-bleeding (Okamura and Ozawa 1995; Rols et al. 1999; Sari et al. 1999; Audenaert et al. 2007; Byung et al. 2007; Maghsoudi and Arabpour-Dahooei 2007; Khanzadi et al. 2010) and compressive strength enhancing additive (Li 2004; Gaitero et al. 2008; Raiess-Ghasemi et al. 2010; Wei and Zhang 2011). Furthermore, the durability and sustainability of concrete structures is becoming of vital importance for the construction industry. In this context, SCC is a type of concrete that has generated tremendous interest since its initial development in Japan by (Okamura and Ozawa 1995).

SCC was developed to obtain durable concrete structures due to an increased homogeneity. For this purpose, a concrete with low yield stress, moderate viscosity and high resistance to segregation, which also can be cast on-site or into precast molds without compaction, was developed. The special flow behavior of SCC is obtained by the use of superplasticizers, high amounts of fine particles, and, in some cases, viscosity modifying agents (added to reduce segregation and bleeding). Due to the presence of high amounts of fines, the pore structure of SCC differs from the pore structure of traditional concrete. According to (Audenaert et al. 2007), the application of SCC might be somewhat risky due to the lack of knowledge concerning the actual durability of this material.

So far, only a few reports on the effects of nano-silica on the durability of SCC are available (Maghsoudi and Arabpour-Dahooei 2007; Raiess-Ghasemi et al. 2010; Wei and Zhang 2011). In addition, the difference in the reactivity of nano-silica due to its production route has not been reported yet.

In this study, three different types of nano-silica were applied in SCC, one produced by the controlled dissolution of the olivine mineral and two having similar particle size distributions (PSD), but produced through two different processes: fumed powder nano-silica and precipitated silica in colloidal suspension. The influence of the nano-silica on SCC was investigated with respect to the properties of the concrete in fresh (workability) and hardened state (durability properties). In addition, the densification of the microstructure of the hardened concrete was analyzed by SEM and EDS techniques.

## 2 Materials and Methods

### *2.1 Materials and SCC Mix Design*

The Portland cement used was CEM I 42.5 N, as classified by (EN 197–1 2000). The coarse aggregates used were composed of broken granite in the fractions 2–8 mm and 8–16 mm. Two different sands were used: dredged river sand (0–4 mm) and microsand (0–1 mm). The microsand is mainly composed of natural sandstone waste that is generated during the crushing process of coarser fractions. A ground limestone powder was applied as filler. Three different nano-silica additives were selected to produce three different SCC batches: one colloidal nano-silica suspension (CnS-2), one fumed powder nano-silica (PnS-3) and one water based suspension with 10 wt.% of nano-silica precipitated from the olivine dissolution (DOnS-8). The two first nano-silica additives (CnS-2 and PnS-3) have similar PSD and specific surface area measured by the BET method (Brunauer et al. 1938), following the standard (DIN-ISO 9277 2005). Furthermore, a superplasticizer (SP1) based on polycarboxylate ethers was added to adjust the workability of the mix. A summary of the general characteristics of all materials used is presented in Table 1. Their PSDs are depicted in Fig. 1.

For the composition of SCC mixes, the mix design concept described in (Hunger 2010) was used. This design concept makes use of an optimization algorithm described in (Hüsken and Brouwers 2008) to compute the mix proportions of all solid ingredients of the concrete mix, following the theory of continuously graded granular mixtures. In the optimization process, a distribution modulus ($q$) of 0.25 was used together with the following constrains: cement content of 340 kg/m$^3$, $w/c$ ratio of 0.45, air content in the fresh mix of 1 % by volume and nano-silica content of 3.8 % bwoc. The cement content and the $w/c$ ratio were selected based on (NEN-EN 206–1 2008) for the exposure class XS3 (aggressive exposure to chlorides originating from seawater). In addition, a flow class of the fresh concrete F7 (630–800 mm) was selected as a target, taking into consideration the Dutch recommendation (BRL 1801 2006) for SCC. An example of the target curve and the composed grading curve of the SCC mix is also shown in Fig. 1.

**Table 1** Properties of the used materials

| Materials | Specific density (g/cm$^3$) | BET (m$^2$/g) | pH | Solid content (% w/w) | Loss on ignition (%) | Computed SSA (m$^2$/m$^3$) |
|---|---|---|---|---|---|---|
| CEM I 42.5 N | 3.14 | 1 | – | – | 2.8 | 1,699,093 |
| CnS-2 | 1.40 | 50 | 9.5 | 50 | – | 46,110,081 |
| PnS-3 | 2.15 | 56 | 5.0[a] | – | 0.5 | 48,175,461 |
| OnS-15 | 2.10 | 400 | 5.8[a] | – | – | 179,142,530 |
| Limestone powder | 2.71 | – | – | – | – | 1,234,362 |
| Microsand (sandstone) | 2.64 | – | – | – | – | 193,514 |
| Sand 0–4 | 2.64 | – | – | – | – | 14,251 |
| Granite 2–8 | 2.65 | – | | – | – | 1,740 |
| Granite 8–16 | 2.65 | – | – | – | – | 515 |
| Superplasticizer | 1.10 | – | 7.0 | 35 | – | – |

[a]4 % w/w in water

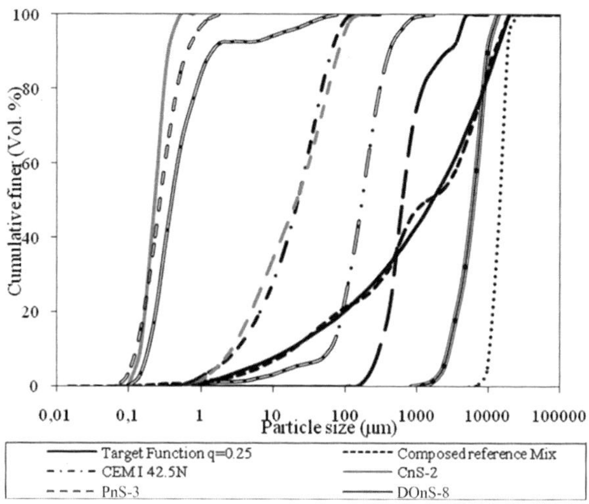


**Fig. 1** PSD of the used materials, target function and composition of reference mix (*dashed line*) for $D_{min} = 60$ nm

## 2.2 Test Methods

### 2.2.1 Fresh Concrete Properties

Using the provided materials, four SCC mixes (two batches of 65 dm$^3$ for each mix) were prepared according to the mix proportions listed in Table 2. These SCCs were mixed for 5 min in total in a compulsory mixer, and subsequently tested for their fresh concrete properties, according to the recommended procedure (EFNARC 2005). The following fresh-concrete properties of the SCC were determined: V-funnel time, slump flow, density, packing density and air content.

**Table 2** SCC concrete mix designed using the optimization tool developed by (Hüsken and Brouwers 2008)

| Materials | Reference | CnS-2 | PnS-3 | DOnS-8 |
|---|---|---|---|---|
| | Concentration in kg/m$^3$ | | | |
| CEM I 42.5 N | 340.0 | 340.0 | 340.1 | 340.1 |
| Nano-silica | 0.0 | 12.8 | 12.8 | 12.8 |
| Limestone powder | 179.4 | 151.8 | 151.9 | 151.8 |
| Microsand (sandstone) 0–1 mm | 125.0 | 141.3 | 141.4 | 141.3 |
| Dredged sand 0–4 mm | 624.3 | 617.9 | 618.0 | 617.8 |
| Granite 2–8 mm | 733.8 | 735.6 | 735.7 | 735.5 |
| Granite 8–16 mm | 274.7 | 274.2 | 274.3 | 274.2 |
| Water | 153.0 | 153.0 | 153.0 | 168.4 |
| Superplasticizer (SP1 PCE-type) | 3.4 | 6.5 | 6.5 | 11.1 |
| Air (Vol. %) | 1.0 | 1.0 | 1.0 | 1.0 |
| Density (g/cm$^3$) | 2.427 | 2.427 | 2.430 | 2.426 |
| $w/c$ | 0.45 | 0.45 | 0.45 | 0.50 |
| $w/p$ | 0.267 | 0.270 | 0.270 | 0.297 |
| Powder content [l/m$^3$] | 194.2 | 192.7 | 192.6 | 192.8 |
| Composed surface (m$^2$/m$^3$) | 277,972 | 547,905 | 554,428 | 940,678 |
| SP1 content (g/m$^2$) | 0.0122 | 0.0119 | 0.0117 | 0.0118 |
| SP1 content (% bwoc) | 1.0 | 1.9 | 1.9 | 3.3 |

*bwoc* based on the weight of cement

### 2.2.2 Hardened Concrete: Durability Tests

Permeable (Water Accessible) Porosity

The permeable porosity affects the transport properties of fluids in concrete and therefore also its durability. It is related to many deterioration processes driven by the transport properties of concrete (Safiuddin and Hearn 2005). In this context, three cores (diameter of 100 mm, height of 150 mm) were drilled with a diamond coring bit from three cubes for each mix. Afterwards, using a diamond saw at medium speed lubricated with water, six concrete discs (height of approximately 15 mm) for each SCC mix were cut from the inner layers of the drilled cores. No polishing of the surface was performed. In total 18 discs were used to determine the permeable porosity, following the procedure described by (ASTM 1202 2005). The vacuum-saturation technique was applied to fill the accessible pores with water, as this technique is concluded to be the most efficient by (Safiuddin and Hearn 2005).

Penetration of Water Under Pressure

The depth of penetration of water under pressure was tested according to (BS-EN 12390-8, 2009) at the age of 28 days. The samples (three cubes for the reference and two for each mix with nano-silica) were exposed to water under pressure (5 bar) for 72 h and subsequently split in order to determine the maximum depth of the water penetration.

### Pore Size Distribution and Porosity of the Paste

The pore size distribution was measured using the mercury intrusion porosimetry (MIP) technique (Autopore IV, Micromeretics). The maximum applied pressure of mercury was 228 MPa, the mercury contact angle was 130° and the equilibrium time was 20 s. The pore size range of 0.0063–900 μm was investigated. Pieces of hardened mortar (approximately 2 g in total) were carefully selected to avoid the presence of coarse sand and gravel. The mortars, extracted from cured SCC samples at 28 days, were first dried in an oven (80 °C) until a constant mass was reached, and then tested by the MIP technique. Only samples from the reference SCC and the SCCs with CnS-2 and PnS-3, respectively, were considered for this study.

### Rapid Chloride Migration Test (RCM)

Two specimens for the RCM test were retrieved from each extracted core giving in total six test specimens (discs, diameter of 100 mm and height of 50 mm) for each mix. Three of these specimens were tested at the age of 28 days and the remaining three at 91 days. One day prior to the RCM test, the specimens were pre-conditioned (vacuum-saturation with limewater). The RCM test was performed according to (NT Build 492 1999), using the test set-up described in (Spiesz et al. 2012). The duration of the RCM test for all samples was 24 h. After the test, the penetration depth of chlorides was measured on split samples by applying a colourimetric indicator for chlorides (0.1 M $AgNO_3$ solution) and the values of the chloride migration coefficients ($D_{RCM}$) were calculated according to (NT Build 492 1999).

### Freeze-Thaw Resistance (Surface Scaling Test)

As a further durability assessment, the freeze-thaw test was performed on SCC samples, even though the air content in the fresh mix was less than the recommended value of 4 % by volume (NEN-EN 206–1 2008). The freeze-thaw resistance, expressed by the surface scaling factor (Sn), was determined following the procedure described in (NEN-EN 12390–9 2006). Nevertheless, the test samples differed from the specifications in the standard. For practical reasons, cylinders were used instead of slabs (Fig. 2a). The 150 mm cubes were cured in water after demolding until the age of 14 days, when the cores (100 mm in diameter) were extracted and sliced (two cylinders of 50 mm in height were obtained from

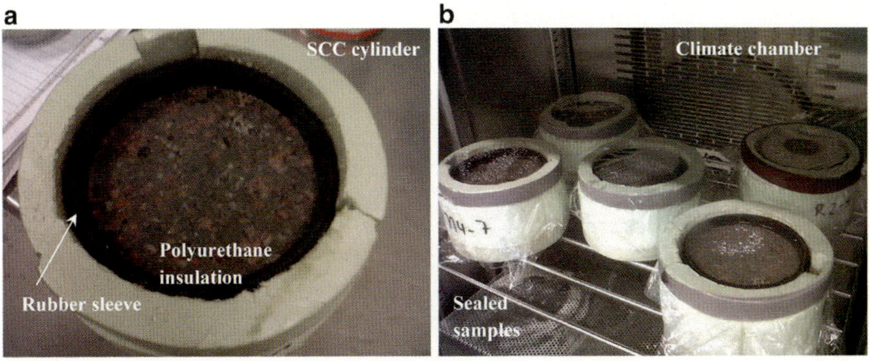


**Fig. 2** Used freeze-thaw experimental setup, (**a**) freeze-thaw samples preparation, (**b**) samples inside the climate chamber after 1 freeze-thaw cycle

each core). Afterwards, the obtained cylinders were cured under water and then sealed after 25 days with tight rubber sleeves. The sealed samples were placed in polyurethane insulations of 10 mm thickness (see Fig. 2a) and surface-saturated with demineralized water for 3 days. Due to a limited volume of the used climate chamber (Fig. 2b), only three specimens were tested for each mix, resulting in a total exposed surface area of 0.024 $m^2$, the area recommended in (NEN-EN 12390–9 2006) is 0.08 $m^2$). After the saturation, the demineralized water was replaced by a 3 mm layer of 3 % by weight NaCl solution, poured on the top surfaces of the samples and then the freeze-thaw cycles were started. The applied temperature profile was following the recommendations given in (NEN-EN 12390–9 2006). The level of solution on the surface of concrete was adjusted regularly. In total, 56 freeze-thaw cycles were performed, during which the surface scaling was measured after 7, 14, 28, 42 and 56 cycles.

### 2.2.3 Microstructural Characterization and Analysis

The microstructural morphology of the prepared concrete was analyzed using a high resolution scanning electron microscope (FEI Quanta 600 FEG-SEM) with a Schottky field emitter gun (at voltage of 10 keV and 0.6 mbar of low-vacuum pressure). Furthermore, a general chemical analysis was performed using EDAX energy dispersive spectroscopy (EDS) detector. Several fracture surfaces were investigated in natural conditions (without any sample preparation). The fracture surfaces of the SCC samples were analyzed in this study to avoid the effect of sample preparation (cutting, grinding, polishing, etc.), which may alter the microstructure of concrete. In the analyses of the fracture surfaces it is possible to characterize the interfacial transition zone (ITZ), the formation of portlandite hexagonal plates and ettringite needles.

# 3 Results and Discussion

## 3.1 Fresh Concrete Properties

The fresh concrete properties of the tested mixes are presented in Table 3. All four mixes fulfilled the requirements for the flow class F7 (630–800 mm of spread diameter), specified in (BRL 1801 2006). Only the mix with CnS-2 resulted in a spread flow close to the lower limit of this target range. Another interesting finding is that the SP content, based on the composed surface area, is nearly constant for all mixes and amounts to an average value of 0.0119 g/m$^2$. This shows that in the present case the required amount of SP can be calculated if the composed surface area of all solid ingredients is known. Nevertheless, the SP dosage depends also on the mineralogy and the surface charge of the fines used (Spiratos et al. 2003; Plank et al. 2009). Another implication of this result is that the SP requirement is rather related to changes in the specific surface area of the mix than to the concentration (by mass) of the nano-silica addition, as it was previously reported by (Sobolev et al. 2006). These researchers concluded that 0.21 % of additional SP is needed for each 1 % of nano-silica added to a standard concrete composition. The SCC with DOnS-8 also fulfils the requirements for the flow class F7.

Another consequence of the high specific surface area (BET of 400 m$^2$/g) and the agglomerated state of the DOnS-8 was the increased water demand. More water was needed to have workability in the specified range (*w*/*c* of 0.50). The increase in the mixing water exceeded the maximum *w*/*c* ratio of 0.45 recommended for the exposure class XS3. Further analysis is performed in the following sections.

Considering the V-funnel time, only the three mixes with nano-silica (CnS-2, PnS-3 and DOnS-8, respectively) fulfil the requirements of the viscosity class VF2 (funnel time 9–25 s) as established by (EFNARC 2005). In this particular case, the SCC with DOnS-8 presented the lowest V-funnel time (18.8 s) mainly caused by its higher amount of added water and SP. Although the reference mix does not fulfill this viscosity class, it is in the range for a high powder content SCC mix, as proposed by (Hunger 2010). In addition, no segregation or blocking was observed

**Table 3** Properties of the prepared SCC mixes in fresh state

| Value | Reference | CnS-2 | PnS-3 | DOnS-8 |
|---|---|---|---|---|
| Slump flow (mm)[a] | 690–720 | 664–701 | 685–720 | 670–720 |
| V-Funnel time (s) | 35.0 | 20.5 | 24.5 | 18.8 |
| Stability time (s) | 69 | 12 | 7.5 | 9.5 |
| Fresh density (g/cm$^3$) | 2.399 | 2.384 | 2.392 | 2.374 |
| Air content (Vol. %)[b] | 1.15 | 1.79 | 1.58 | 2.17 |
| Packing density (%)[b] | 83.55 | 82.91 | 83.12 | 80.99 |

[a]Maximum and minimum flow diameter measured
[b]Calculated value

for any mixes. A long V-funnel time was already reported as a typical characteristic of SCC with increased amount of limestone powder (Kordts and Breit 2004; Felekoglu 2007).

As also can be seen in Table 3, the SCC mixes with nano-silica have higher air content compared to the reference mix. In this case, SCC with DOnS-8 has the higher air content of all mixes (2.17 %), which is caused by the higher viscosity of the paste due to the application of particles with high specific surface area (400 vs. 50 $m^2/g$). The air entrainment was also confirmed by the difference between the designed and the measured concrete density (see Table 3).

## 3.2 *Hardened Concrete: Durability*

### 3.2.1 Permeable Porosity

The results of the measurements of the permeable porosity of the SCC mixes are presented in Table 4. These results are surprisingly showing that the SCC reference mix has slightly lower porosity (12.1 %) compared to the mixes containing the three types of nano-silica (12.5–13.1 %). The porosity and the tortuosity of the pores in the hardened cement paste are normally reduced when pozzolanic materials are added, and this influences many properties such as the compressive and splitting tensile strength (Garboczi 1990). Nevertheless, in (Yogendran and Langan 1987) it is stated that for the addition of micro-silica the total pore volume is not necessarily changed, but larger pores appear to be subdivided into smaller pores. Apparently, the same behavior was found for the SCC in the present case. The air content of the fresh mix influenced, probably, the final porosity. Some researchers (Safiuddin and Hearn 2005) state that the vacuum-saturation technique is also able to take void pores into account. This means that the porosity values shown in Table 4 consider also the differences in the air content of the mixes (the lower the air content the lower the permeable porosity). Also, it is known that a higher amount of mixing water will produce higher capillary pores in the set concretes (Neville 2002), which is the case of the SCC with DOnS-8 additions.

**Table 4** Twenty-eight-day permeable porosities and penetration depths of water under pressure of the three designed SCC mixes

| SCC properties | Reference | CnS-2 | PnS-3 | DOnS-8 |
|---|---|---|---|---|
| Permeable porosity (Vol. %) | $12.07 \pm 0.17$ | $12.45 \pm 0.26$ | $12.48 \pm 0.23$ | $13.06 \pm 0.38$ |
| Penetration depth of WUP[a] (mm) | $26 \pm 7$ | $3 \pm 2$ | $3 \pm 2$ | $10 \pm 3$ |

[a]WUO: water under pressure (5 bar)

### 3.2.2 Penetration of Water Under Pressure

The test results of the penetration of water under pressure are shown in Table 4. According to (Raiess-Ghasemi et al. 2010), all tested samples are in the low permeability range (penetration depth of less than 30 mm). Additionally, the two SCC with CnS-2 and PnS-3 types of nano-silica presented penetration depths less than 5 mm. The SCC with DOnS-8 resulted in a penetration depth of 10 mm. This means that despite the higher w/c ratio used and total permeable porosity measured the addition of olivine nano-silica improved or modified the microstructure of the pores, probably changing its tortuosity or shapes. This implies, in general, that the addition of 3.8 % nano-silica results in concrete which is highly resistant to the penetration of water under pressure of 0.5 MPa (5 bar). Similar permeability improvements were reported by (Raiess-Ghasemi et al. 2010) for conventional concrete with micro and nano-silica addition. The results also suggest that, despite the similar permeable porosity of all SCC mixes, the nano-silica samples have very low effective water permeability (less interconnected pores and/or finer pore structure).

### 3.2.3 Pore Size Distribution and Porosity of the Concrete Matrix

To support the findings hardened mortar (matrix) of the concrete samples were analyzed using mercury intrusion porosimetry (MIP). The obtained results are shown in Figs. 3 and 4 for the reference, CnS-2 and PnS-3 SCCs, respectively. In addition, the properties of the hardened matrix (mortar) of SCC mixes obtained from MIP measurements are shown in Table 5. In Figs. 3 and 4 one can notice that the addition of 3.8 % of nano-silica slightly increases the volume of pores smaller than 20 nm (indicated with black arrows in Figs. 3 and 4). Similar trends were obtained for the overall parameters that were extracted from the mercury intrusion test.

It is clearly shown that the addition of nano-silica decreased the median pore diameter (by volume and area) and the average pore diameter, which is reduced from 27.8 nm (for the reference) to 24.3 and 25.9 nm for the colloidal nano-silica and powder nano-silica, respectively. Apparently, a reduction of 4 nm in the average pore diameter was enough to significantly reduce the permeability of the SCC containing nano-silica. This means that the modification of the SCC pore structure due to nano-silica addition is reflected in changes in the medium capillary pores and the gel pores (the minimum pore size detected with MIP was 8 nm). Permeability and penetration of harmful substances into concrete are affected mainly by the large and medium capillary pores (Neville 2002). Apparently, nano-silica made the pore structure of paste more homogeneous by increasing medium capillary porosities. In a similar way, the gel pores volume was increased (see Fig. 4). The increased gel porosity can be caused by an increased amount of CSH-gel in the paste, due to an acceleration effect of the nano-silica. The increased gel porosity could result from a higher hydration degree and from the properties of the gel produced by the pozzolanic

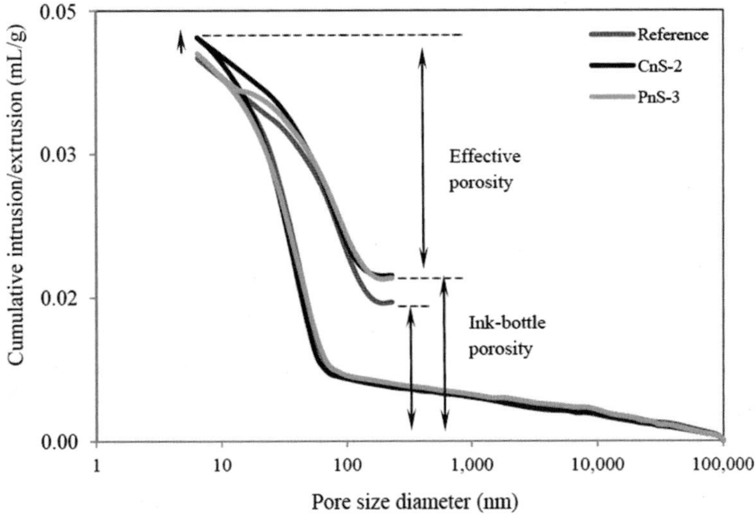


**Fig. 3** Cumulative intrusion/extrusion vs. pore size curves of hardened paste extracted from the tested SCC mixes. *Black arrows* in the figures indicate changes or displacement of the curves due to the addition of nano-silica

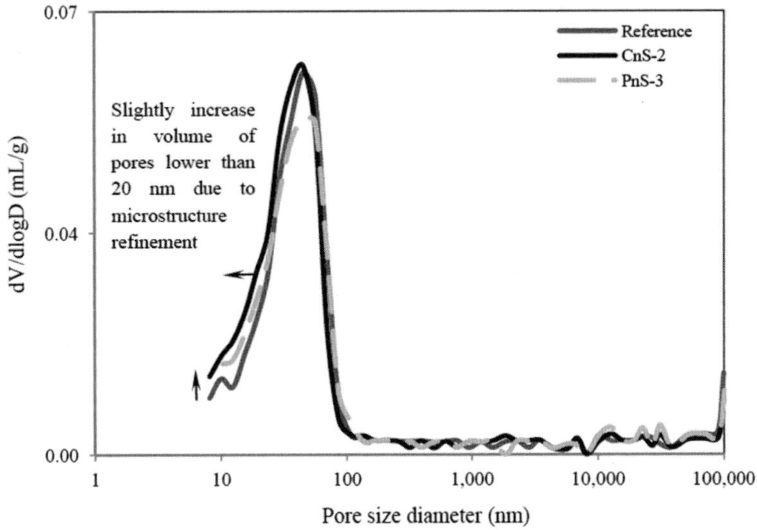


**Fig. 4** Log differential intrusion vs. pore size curves of hardened paste extracted from the tested SCC mixes. *Black arrows* in the figures indicate changes or displacement of the curves due to the addition of nano-silica

**Table 5** Properties of the hardened matrix (mortar) of SCC mixes obtained from MIP measurements

| Properties | Reference | CnS-2 | PnS-3 |
|---|---|---|---|
| Median pore diameter by volume (nm) | 34.8 | 30.9 | 33.8 |
| Median pore diameter by area (nm) | 20.3 | 17.2 | 17.5 |
| Average pore diameter, $4V/A^{+}$ (nm) | 27.8 | 24.3 | 25.9 |
| Bulk density at 0.1 MPa (g/ml) | 2.216 | 2.222 | 2.239 |
| Apparent skeletal density* (g/ml) | 2.415 | 2.438 | 2.445 |
| Porosity (%) | 8.79 | 9.31 | 8.99 |

(*): Determined excluding the Hg accessible pores, (+): ratio between four times median pore diameter by volume and the median pore diameter by area (4V/A)

reaction of the nano-silica particles. Pozzolanic CSH-gels, normally, have a lower Ca/Si (1–0.8) ratio and different gel porosities (Lothenbach et al. 2011).

Isolated (non-accessible) pores do not contribute to the transport properties of fluids and cannot be detected by MIP. Nevertheless, taking into account the differences between the intrusion and the extrusion cycles of mercury (Fig. 3) and the definition of effective porosity and ink-bottle porosity defined in (Ye 2003), it is possible to observe that the addition of nano-silica increased the amount (volume) of ink-bottle pores. The increment is around 7 % (+0.0028 mL/g) for both types of nano-silicas. The transport of species through the ink-bottle pores is more difficult due to their constrictivity. These differences can explain the changes observed in the pores structure of the cement pastes and can partially explain the results of water pressure penetration test, including the WUP results obtained with DOnS-8. Even though, the MIP analysis performed gives some evidence of the changes in the concrete pore structure, the results are close to the test repeatability. For that reason, complementary pore size distribution analyses and other characterization techniques enabling to detect changes in the pore structure are recommended in further research.

### 3.2.4 Rapid Chloride Migration Test (RCM)

The Rapid Chloride Migration test, performed according to (NT Build 492 1999), is a commonly used accelerated technique for determining the chloride transport rate in concrete. The output of the test (the so-called chloride migration coefficient $D_{RCM}$) is regularly employed in service-life design models for concrete elements and structures exposed to chlorides. In this context, Fig. 5 presents the average values of the calculated chloride migration coefficients ($D_{RCM}$) of each SCC mix studied.

Like the conductivity test results, the migration coefficients are much lower for the mixes containing nano-silica. Additionally, the SCC mix with CnS-2 shows again the best performance. On the contrary, the SCC mix with DOnS-8 showed similar performance than the reference mix even thought its higher w/c ratio. The explanation of this behavior is the same as previously discussed a finer

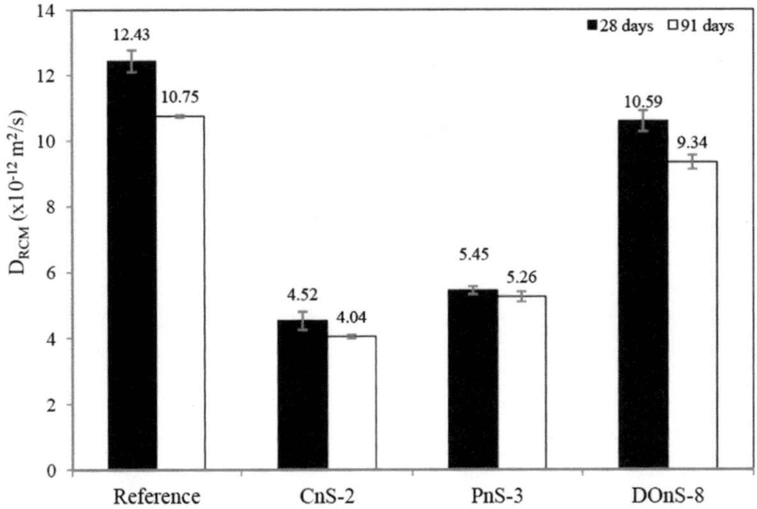


**Fig. 5** Twenty-eight and ninety-one days SCC chloride migration coefficient (DRCM)

porosity, greater tortuosity and more precipitated CSH-gel decrease the mobility of the chloride ions into the SCC pore solution. It can also be noticed that the 91 days $D_{RCM}$ values are lower compared to the 28 days $D_{RCM}$. This can be explained by the progressing hydration of cement and a related densification of the microstructure of concrete.

### 3.2.5 Freeze-Thaw Resistance

The results of the freeze-thaw surface scaling of the four selected SCC mixes are shown in Fig. 6. The failure of the SCC reference mix, taking into account the maximum scaling criteria of 1.5 kg/m$^2$ at 28 cycles as recommended in (Romero et al. 2011), occurred about the 11th cycle. On the other hand, the SCC mixes with CnS-2 and PnS-3 addition resulted in a surface scaling factor lower than the recommended value for non-air entrained concrete after 28 cycles. These SCC mixes failed the 1.5 kg/m$^2$ criterion after 52 cycles. In contrast, the SCC with DOnS-8 additions showed a scaling factor lower than the reference (21 cycles) but higher than the specified limit for non-air entrained concrete until 28 cycles. Better resistance to the freeze-thaw cycles of the SCC with DOnS-8 additions can be attributed to its denser microstructure and higher air content compared to the reference (see Table 2). The highly stiff CSH-gel and the refined pore structure results in a limited intrusion of water and in an improved resistance to changes of temperature in the concrete surface.

The freeze-thaw resistance depends on the compressive strength, porosity, air void content and other parameters such as the air-voids distribution and pore size (Neville 2002). A better resistance to the freeze-thaw induced damage of the SCC

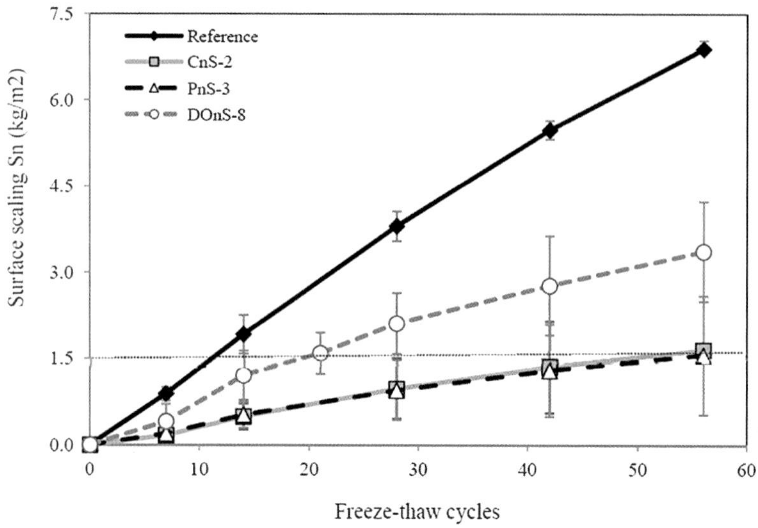


**Fig. 6** Cumulative scaling factor (Sn) after 56 freeze–thaw cycles (3 % NaCl solution)

with nano-silica addition can be attributed to different factors, such as the formation of highly stiff CSH-gel with E modules of 26 GPa reported by (Mondal et al. 2010) and the refined pore structure (higher tortuosity and constrictivity of the pores), which results in a limited intrusion of water and in improved resistance to the changes of temperature near the surface of the concrete. Despite the better freeze-thaw resistance of the SCC with nano-silica compared to the reference mix, its scaling values are larger than the recommended value of 0.5 kg/m$^2$ after 56 cycles, suggested by (Stark and Wicht 2001) for a concrete classified as having good resistance against freeze-thaw exposure. Nevertheless, with an air entrainment admixture that guarantees a minimum air content of 4 %, the freeze-thaw resistance of SCC with nano-silica should result in a mix having a high resistance to freeze-thaw.

## 3.3 *Microstructural Analysis*

The objective of the microstructural analysis is to support the findings shown in the present research. It is important to notice, that the performed microstructural analyses were qualitative, with the objective to give additional information that can explain the results obtained in the other executed tests.

The SCC with colloidal nano-silica shows a more homogeneous microstructure compared to the reference mix. As an example, Fig. 7a shows the results obtained for CnS-2 SCC. A homogenous microstructure is defined by a more regular morphology and similar size of the hydrates. This microstructure is characterized

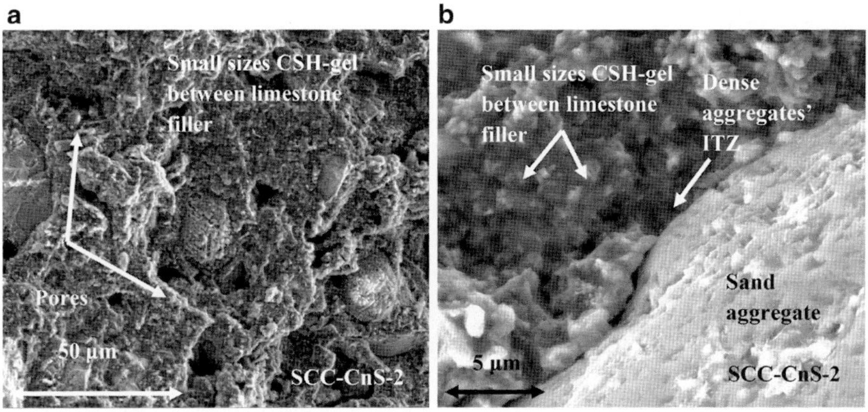


**Fig. 7** Microstructure low vacuum (0.6 mbar) FEG–SEM photomicrographs of the SCC-CnS-2 mix, (**a**) cement matrix, (**b**) aggregate dense ITZ

by apparent compact and small-sized C-S-H gel and the absence of abundant gel porosity. As a consequence, a relatively denser ITZ was also confirmed by SEM (Fig. 7b). It is important to notice, that the needle-type structures or well grown Ca $(OH)_2$ crystals were not found in the microstructural analysis. The absence of needle-type hydrates in the cement matrix is unexpected. In general, as was explained for the reference SCC, ettringite is one of the hydrated phases formed in cement system with high amount of limestone additions. In the case of the SCC with CnS-2 the concentration of limestone was slightly decreased (from 179.4 to 151.8 kg/m$^3$), but this decrease should not influence the hydrates phases formation. Similarly, in system with amorphous nano-silica it was predicted (Lothenbach et al. 2011) that one of the stable phases formed is ettringite. Ettringite in long-term is only unstable when the pH of the pore solution falls below 10 due to further replacement of cement by silica. The dropping of the pore solution pH to values lower than 10 is predicted for amorphous silica additions higher than 35 % by mass (Lothenbach et al. 2011). In the present research only 3.8 % (based on the weight of cement) of nano-silica is added. One possibility that can explain the absence of ettringite is that the addition of nano-silica decreased the size of ettringite crystals, making much more difficult to identify by the used magnification of the ESEM. Also, smaller ettringite crystals are more susceptible to be discomposed by drying and vacuum conditions as was reported by (Taylor 1990). As was mentioned before, well grown $Ca(OH)_2$ crystals were not found in the microstructural analysis. Changes in the amount and size of portlandite crystals are expected in the presence of amorphous silica as was established by different authors (Rols et al. 1999; Ji 2005; Sobolev and Ferrara 2005; Sobolev et al. 2006; Lin et al. 2008; Gaitero et al. 2008; Senff et al. 2009; Belkowitz and Armentrout 2010). The absence of well crystallized portlandite and the lower amounts of gel pores confirm that the addition of the nano-silica causes a refinement of the microstructure and probably induces the precipitation of small-sized CSH-gel that normally has a high stiffness and lower Ca/Si ratio (Mondal et al. 2010).

## 4 Conclusions

Based on the workability and durability properties obtained by the tests on a reference SCC and SCC mixes containing three types of nano-silica, the following conclusions can be drawn:

1. The results of the fresh state behavior of SCC demonstrated that concrete with an addition of 3.8 % nano-silica (based on the mass of cement) shows similar flow properties and viscosity as the reference mix without nano-silica. Slightly increased air content in the mixes with nano-silica was measured due to a higher viscosity of the paste.
2. The water permeable porosity of the three SCC mixes was found to be similar. On the other hand, when nano-silica (CnS-2 and PnS-3 type) were added, the concrete becomes almost impermeable to the penetration of water under pressure. This can be explained by a refinement of the microstructure of the paste due to higher amount of pores with ink-bottle shape, as confirmed by MIP measurements.
3. The results of the fresh state behavior of SCC with olivine nano-silica (DOnS-8) demonstrated that concrete with addition of 3.8 % based on the mass of cement shows better flowing and viscosity behavior than the reference mix. The high reactivity of the olivine nano-silica increased the water demand and resulted in higher air content of the mix. The compressive and tensile splitting strength of SCC with DOnS-8 was not improved due to the higher $w/c$ ratio. On the contrary, the water permeable porosity was similar to the reference mix. Nevertheless, when olivine nano-silica is added, the SCC mixes reduced their permeability to the penetration of water under pressure.
4. All durability indicators of the SCC studied (conductivity, chloride migration coefficients, and freeze-thaw resistance) were significantly improved with the addition of 3.8 % of the three types of the nano-silica. Moreover, the SCC with colloidal nano-silica showed slightly better properties than the SCC with powder nano-silica.
5. The microstructural analysis of the hardened SCC reveals that, the addition of nano-silica resulted in a homogeneous microstructure, characterized by compact and small-sized CSH-gel. As a consequence, a denser ITZ was produced. The addition of nano-silica caused a refinement of the microstructure (less interconnected and finer pore structure) and induced the precipitation of small-sized CSH-gel, probably having a higher stiffness and lower Ca/Si ratio.
6. The improvement of the microstructure resulted in a significant reduction of the permeability of concrete (good resistance against the penetration of chlorides and water under pressure) because of the microstructural densification and increased tortuosity of the pore system.
7. The high reactivity and faster pozzolanic behavior of the colloidal nano-silica (CnS-2) particles at early age produced a more refined microstructure than obtained for the SCC with powder nano-silica and olivine nano-silica.

Nevertheless, the higher reactivity and agglomerated state appears to be the reason for the higher deviation standard obtained in the compressive strength results.

**Acknowledgements** This research was carried out under the project number M81.1.09338 in the framework of the Research Program of the Materials innovation institute (M2i) and The European Community's Seventh Framework Program, ProMine: Nano-particle products from new mineral resources in Europe, FP7-NMP-2008-LARGE-2 under grant agreement 228559. Furthermore, the authors thank ir. J.J.W. Gulikers for his help and assistance.

## References

ASTM C1202 (2005) Standard test method for electrical indication of concrete's ability to resist chloride ion penetration. In: Annual book of ASTM standards, vol 04.02. American Society for Testing and Materials, Philadelphia, pp 1–6

Audenaert K, Boel V, De Schutter G (2007) Chloride migration in self compacting concrete. In: Proceedings of the fifth international conference on concrete under severe conditions: environment and loading (CONSEC'07), Tours, 4–6 June 2007, pp 191–298

Belkowitz JS, Armentrout D (2010) An investigation of nano-silica in the cement hydration process. In: Proceedings of the 2010 concrete sustainability conference, National Ready Mixed Concrete Association, Silver Spring, pp 1–15

BMC Certificatie BRL 1801 (2006) Nationale beoordelingsrichtlijn betonmortel. BMC Certificatie, Gouda (in Dutch)

Brunauer S, Emmet PH, Teller E (1938) Adsorption of gases in multimolecular layers. J Am Chem Soc 62:309–319

BS-EN 12390-8 (2009) Testing hardened concrete – depth of penetration of water under pressure. British Standards Institution-BSI and CEN European Committee for Standardization, pp 1–10

Byung WJ, Chang HK, Jae HL (2007) Investigations on the development of powder concrete with nano-$SiO_2$ particles. KSCE J Civil Eng 11(1):37–42

CEN EN 197–1 (2000) Cement-part 1: composition and specifications and conformity criteria for common cements, European Commission for Standardization (CEN), pp 1–33

DIN ISO 9277 (2005) Determination of the specific surface area of solids by gas adsorption using the BET method. German Institute of Normalization-DIN, pp 1–19

EFNARC (2005) Specification and guidelines for self compacting concrete-SCC, Report, European Federation of Producers and Contractors of Specialist Products for Structures, Surrey, UK

Felekoglu B (2007) Utilization of high volumes of limestone quarry wastes in concrete industry (self-compacting concrete case). Resour Conserv Recycl 51:770–791

Gaitero JJ, Campillo I, Guerrero A (2008) Reduction of the calcium leaching rate of cement paste by addition of silica nanoparticles. Cem Concr Res 38:1112–1118

Garboczi EJ (1990) Permeability, diffusivity, and microstructural parameters: a critical review. Cem Concr Res 20:591–601

Hunger M (2010) An integral design concept for ecological self-compacting concrete. PhD thesis, Eindhoven University of Technology

Hüsken G, Brouwers HJH (2008) A new mix design concept for earth-moist concrete: a theoretical and experimental study. Cem Concr Res 38:1246–1259

Iler RK (1955) The colloid chemistry of silica and silicate. Cornell University Press, Ithaca, pp 1–250

Ji T (2005) Preliminary study on the water permeability and microstructure of concrete incorporating nano-$SiO_2$. Cement Concr Res 35:1943–1947

Khanzadi M, Tadayon M, Sepehri H, Sepehri M (2010) Influence of nano-silica particles on mechanical properties and permeability of concrete. In: Proceedings of the second international conference on sustainable construction materials and technologies, Universita Ploitecnica delle Marche, Ancona, 28–30 June 2010, pp 1–7

Kordts S, Breit W (2004–2006) Combined test method for assessing the workability of SCC-Flow cone, Concrete Technology reports, pp 7–15

Li G (2004) Properties of high-volume fly ash concrete incorporating nano-$SiO_2$. Cem Concr Res 34:1043–1049

Lin KL, Chang WC, Lin DF, Luo HL, Tsai MC (2008) Effects of nano-$SiO_2$ and different ash particle sizes on sludge ash–cement mortar. J Environ Manage 88:708–714

Lothenbach B, Scrivener K, Hooton RD (2011) Supplementary cementitious materials. Cem Concr Res 41:1244–1256

Maghsoudi AA, Arabpour-Dahooei F (2007) Effect of nanoscale materials in engineering properties of performance self compacting concrete. In: Proceeding of the 7th international congress on civil engineering, Iran, pp 1–11

Mondal P, Shah SP, Marks LD, Gaitero JJ (2010) Comparative study of the effects of microsilica and nanosilica in concrete. In: Transportation research record: journal of the transportation research board, vol 2141. Transportation Research Board of the National Academies, Washington, DC, pp 6–9

NEN-EN 12390–9 (2006) Testing hardened concrete – freeze-thaw resistance – scaling. CEN European Committee for Standardization and Dutch Normalization-Institute, Delft (in English), pp 1–29

NEN-EN 206–1 (2008) Beton Deel 1: Specificatie, eigenschappen, vervaardiging en conformiteit. Nederlands Normalisatie Instituut, Delft, pp 1–72

Neville AM (2002) Properties of concrete, 4th edn. Prentice Hall/Pearson, Harlow, pp 537–576

Nili M, Ehsani A, Shabani K (2010) Influence of nano-SiO2 and micro-silica on concrete performance. In: Proceedings of the second international conference on sustainable construction materials and technologies, Universita Ploitecnica delle Marche, Ancona, Italy

Nordtest method NT Build 492 (1999) Concrete, mortar and cement-based repair materials: chloride migration coefficient from non-steady-state migration experiments. NT BUILD 492, Esbo, pp 1–8

Okamura H, Ozawa K (1995) Mix-design for self-compacting concrete. Concr Libr Jpn Soc Civil Eng JSCE 25:107–120

Plank J, Schroefl CH, Gruber M, Lesti M, Sieber R (2009) Effectiveness of polycarboxylate superplasticizer in ultra-high strength concrete: the importance of PCE compatibility with silica fume. J Adv Concr Technol 7(1):5–12, Japan

Quercia G, Brouwers HJH (2010) Application of nano-silica (nS) in concrete mixtures. In Fisher G, Geiker M, Hededal O, Ottosen L, Stang H (eds) The 8th fib international PhD symposium in civil engineering, Lyngby, 20–23 June, pp 431–436

Raiess-Ghasemi AM, Parhizkar T, Ramezanianpour AA (2010) Influence of colloidal nano-SiO2 addition as silica fume replacement material in properties of concrete. In: Proceedings of the second international conference on sustainable construction materials and technologies, Universita Ploitecnica delle Marche, Ancona, 28–30 June 2010, pp 1–8

Rols S, Ambroise J, Péra J (1999) Effects of different viscosity agents on the properties of self-leveling concrete. Cem Concr Res 29:261–266

Romero HL, Casati MJ, Galvez JC, Molero M, Segura I (2011) Study of the damage evolution of concrete under freeze-thaw cycles using traditional and non-traditional techniques. In: Proceeding of the XIII international conference on cement chemistry, Madrid, 4–8 July 2011, pp 1–7

Safiuddin M, Hearn N (2005) Comparison of ASTM saturation techniques for measuring the permeable porosity of concrete. Cem Concr Res 35:1008–1013

Sanchez F, Sobolev K (2010) Nanotechnology in concrete – a review. Constr Build Mater 24:2060–2071

Sari M, Prat E, Labastire JF (1999) High strength self-compacting concrete original solutions associating organic and inorganic admixtures. Cem Concr Res 29:813–818

Scrivener KL, Kirkpatrick RJ (2008) Innovation in use and research on cementitious material. Cem Concr Res 38:128–136

Senff L, Labrincha JA, Ferreira VM, Hotza D, Repette WL (2009) Effect of nano-silica on rheology and fresh properties of cement pastes and mortars. Construct Build Mater 23:2487–2491

Sobolev K, Ferrara M (2005) How nanotechnology can change the concrete world – Part 1. Am Ceram Bull 84(10):14–17

Sobolev K, Flores I, Hermosillo R (2006) Nanomaterials and nanotechnology for high-performance cement composites. In: Proceedings of ACI session on nanotechnology of concrete: recent developments and future perspectives, Denver, 7 Nov 2006, pp 91–118

Spiesz P, Ballari MM, Brouwers HJH (2012) RCM: a new model accounting for the non-linear chloride binding isotherm and the non-equilibrium conditions between the free- and bound-chloride concentrations. Construct Build Mater 27:293–304

Spiratos N, Page M, Mailvaganam NP, Malhotra VM, Jolicoeur C (2003) Superplasticizer for concrete fundamentals, technology, and practice. Supplementary Cementing Materials for Sustainable Development Inc., Ottawa, pp 1–321

Stark J, Wicht B (2001) Dauerhaftigkeit von Beton: Der Baustoff als Werkstoff, Birkhäuser, Basel (in German)

Taylor HFW (1990) Cement chemistry. Academic, London

Wei X, Zhang P (2011) Sensitivity analysis for durability of high performance concrete containing nanoparticles based on grey relational grade. Mod Appl Sci 5(4):68–73

Ye G (2003) Experimental study and numerical simulation of the development of the microstructure and permeability of cementitious materials. PhD thesis, Delft University of Technology, Delft, 18 Dec 2003, pp 1–189

Yogendran V, Langan BW (1987) Utilization of silica fume in high strength concrete. In: Proceedings of utilization of high strength concrete, Stavanger, Tapir Publisher, Trondheim

Zhang M, Ilsam J, Peethamparan S (2012) Use of nano-silica to increase early strength and reduce setting time of concretes with high volumes of slag. Cem Concr Compos 34:650–662

# The Effect of Climate Change on Freeze-Thaw Cycles in Nordic Climate

**Toni A. Pakkala, Arto Köliö, Jukka Lahdensivu, and Matti Pentti**

**Abstract** In Finland weather conditions are very aggressive for concrete structures which are not sheltered against rain. One of the most serious issues concerns freeze-thaw cycles when at the same time ambient conditions are wet. This situation is common in Finland during autumn, spring and warm winter when daytime temperature is slightly above 0 °C and by night below.

Based on earlier studies the number of freeze-thaw cycles is slightly higher inland than in Southern Finland and coastal areas but the higher amount of rain and sleet in coastal areas makes the number of freeze-thaw cycles needed for the same degree of frost damage much lower. This paper studies how the change of freeze-thaw cycle occurrence affects durability of concrete structures. The study utilizes also models the Finnish Meteorological Institute (FMI) has developed for climate change effects on critical climatic conditions based on three IPCC (2007) scenarios for the evolution of greenhouse gas.

**Keywords** Concrete structure • Frost attack • Climate change • Durability

## 1 Introduction

### *1.1 Finnish Building Stock*

The Finnish building stock consists of 2.4 million buildings which include over 1.2 million residential buildings. Most of the residential buildings are detached houses in which almost half of the Finnish population lives. Apartment houses are concentrated in big cities and 34 % of the population lives in these housings (RIL 2013).

The volume of the renovation is increasing because the Finnish building stock is homogenous and fairly new compared to e.g. South and Middle Europe. Most of the stock is built after the 1960s. For example, construction with precast concrete panels increased at that time because of the urbanization and most of it is built

T.A. Pakkala (✉) • A. Köliö • J. Lahdensivu • M. Pentti
Department of Civil Engineering, Tampere University of Technology, Tampere, Finland
e-mail: toni.pakkala@tut.fi



C. Andrade et al. (eds.), *Durability of Reinforced Concrete from Composition to Protection*, DOI 10.1007/978-3-319-09921-7_13

during the 1960s and 1970s. Buildings constructed at that time are now coming at the end of their service life and thus they need to be renovated. Although the share of precast concrete panels and other concrete facades is only 18 % of all the facades its renovation volume is and will become significant within the near future (Lahdensivu 2012).

## 1.2 Prefabricated Concrete Structures

Since 1970s almost all prefabricated concrete structures in Finland are based on the Concrete Element System (BES 1969) which defines, for instance, the recommended floor-to-floor height and the types of prefabricated panels used. The concrete facade panels used in exterior walls of multi-storey residential buildings were, and still are, predominantly prefabricated sandwich-type panels with thermal insulation placed between two concrete layers, see Fig. 1. It should be noted that usually there is no ventilation gap behind the outer layers of precast exterior wall panels. Thus, if the thermal insulation gets wet e.g. due to leakage through the joints, the structure will dry quite slowly because the drying is a consequence of thermal conduction through the structure from inside to outside. The drying of the outer layers is also slow because of the efficient thermal insulation that limits the drying heat flow through the wall. This means that the concrete may remain moist for long periods.

On the other hand, the most common balcony type in Finland from the late 1960s until today consists of a floor slab, side panels and a parapet panel of precast concrete. These stacked balconies have their own foundations, and the whole stack is connected to the building frame only to brace it against horizontal loads. All structural members of a precast balcony are load-bearing. The cross-section of a typical balcony constructed of precast panels is presented in Fig. 2.

The frost resistance of concrete used in facades or balconies will be determined, in the first place, during concrete mixing process. Air entrainment to fresh concrete

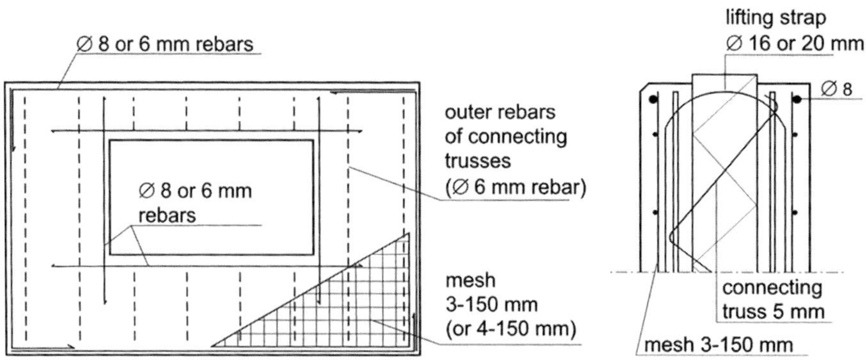


**Fig. 1** Typical reinforcement of outer layer of a concrete facade panel (Pentti et al. 1998)

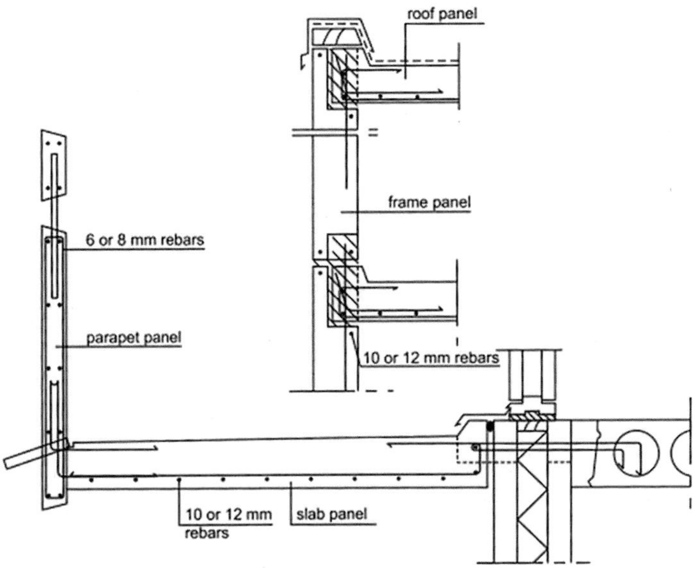


**Fig. 2** Cross-section of a typical finnish balcony made of precast structural members (Pentti 1994)

has been recommended since 1976 and demanded since 1980 in Finnish concrete codes. The compressive strength of concrete has been increased to present level (C30/37) as early as 1989 (BY32 1989).

## *1.3 Climate in Finland*

Although the Finnish climate is relatively steady for the latitudes it still varies significantly from mild and relatively rainy coastal climate to drier continental climate in inland. However, the Finnish building stock is mainly focused on the biggest cities and certain growth areas near them. Due to both the climate differences and the concentration areas of the population Finland can be divided into four main areas: coastal area, southern Finland, inland and northern Finland (Lapland region). The coastal area includes a 30 km wide sector along the coast from the city of Vaasa to the Russian border. Southern Finland includes the rest of the southern parts to the level of the Tampere city (150 km north from Helsinki). Inland area includes the rest of the country except Lapland, see Fig. 3.

Prevailing wind directions and wind speeds also have a strong influence on the distribution of rainfall across a building. In Finland most of the rain and sleet in wintertime comes with southerly to westerly winds. Rain events with wind from other directions have been rare. Due to stronger winds, about 60 % of the rain and sleet load in the coastal area hits the facades and balconies; the corresponding share in the inland is about 40 %. Combined with the higher amount of precipitation in

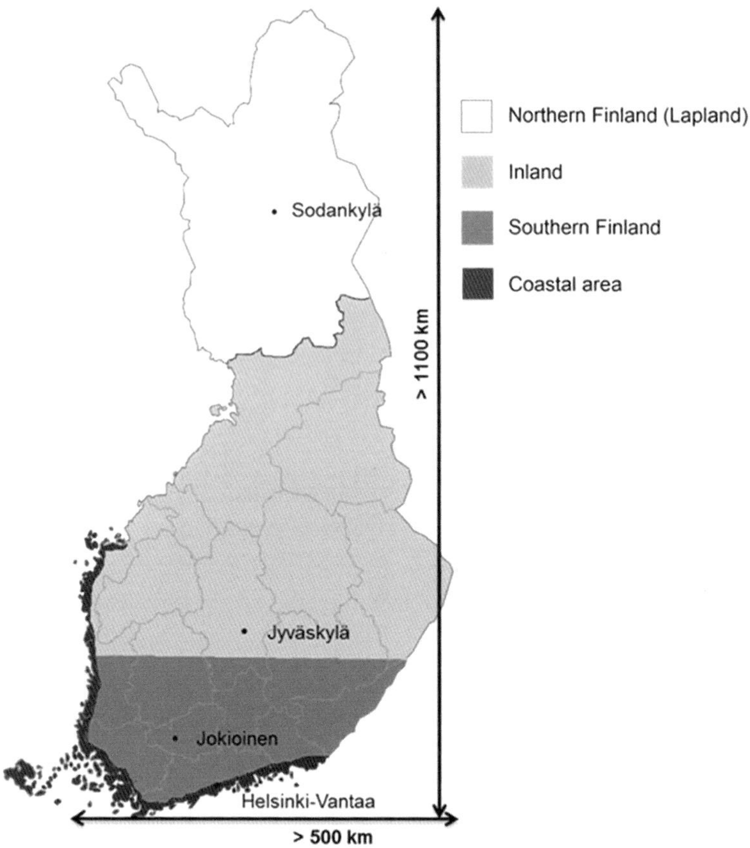


**Fig. 3** Finland can be divided to four main areas due both climate differences and concentration areas

coastal areas, facades and balconies there are subject to considerably higher moisture load than inland resulting in clearly more corrosion and frost damage. Winds are stronger at higher levels of buildings than close to ground level which naturally leads to upper sections of high buildings receiving more rain and sleet stress than lower buildings, and lower sections of buildings in general (Lahdensivu et al. 2013).

## 1.4 *Frost Attack on Concrete*

Frost attack due to a high moisture load is a common reason for the deterioration of concrete structures in Nordic outdoor climate. Concrete is a porous material and its pore system may, depending on the conditions, hold varying amounts of water. As the water in the pore system freezes, it expands about 9 % by volume creating hydraulic pressure in the system. If the level of water saturation of the system is high, the overpressure cannot escape into air-filled pores and thus will damage the internal structure of the concrete resulting in its degradation (Pigeon and Pleau 1995).

Prerequisites for frost damage are that pore structure of concrete has been capillary fulfilled over critical point (Fagerlund 1977) and freezing of pore water will happen in the temperatures low enough (Pigeon and Pleau 1995). Thus, even if concrete is not frost resistant there will not occur any frost damage if concrete is dry or temperature does not go low enough. In practice, frost damage needs the concrete to freeze to temperatures of −5 °C or lower.

The early phase of frost damage is manifested as inner cracking of concrete or scaling of the concrete surface when the hydraulic pressure caused by freezing pore water exceeds the tensile strength of concrete. Cracking will decrease the strength of concrete and increase capillary water absorption. Continuing freeze-thaw cycles and a high moisture content of concrete will finally lead to frost damage (Neville 1995). Early phase frost damage is not visible and cannot be detected by hammering the surface of concrete. Detection of such inner cracking of concrete requires a more sophisticated research method like thin-section analysis (Pentti et al. 1998). Far advanced frost damage is manifested as a reduction in strength of concrete, loss of adhesion, or crazing or chipping off of the surface due to internal expansion. Disintegration of concrete also accelerates carbonation of concrete due to cracking and consequently also steel corrosion.

### *1.5 Objectives*

The main objective of this study is to find out how the climate change effect on the amount of freeze-thaw cycles will affect frost durability properties of outdoor concrete structures.

## 2 Research Material and Methods

### *2.1 Concrete Facades and Balconies Database*

Research material for this study is composed of data on concrete durability collected from condition assessments conducted on prefabricated concrete facades and balconies. As this study discusses the current concrete codes, only the data concerning buildings built in 1990 or after are taken into consideration. In all there are 72 buildings from this era in the database. The database contains measurements of concrete pore structure, tensile strength etc. In addition it includes results from thin section analyses and visual observations made from the building facades and balconies.

## 2.2 Climate Data and Projections

The Finnish Meteorological Institute (FMI) has weather observation data since 1961 in digital form from several meteorological stations covering all of Finland. The data consist of temperature, relative humidity, rain intensity, wind speed and direction, solar radiation variables, etc. These observations have been collected at least daily and three times a day at best.

In the ACCLIM project (Jylhä et al. 2009) the FMI examined the different climate models, and thereupon developed models for Finnish climate conditions and forecasted future climate conditions based on three IPCC (2007) scenarios for the evolution of greenhouse gas and aerosol particle emissions. Based on the scenarios the FMI has presented their effects on weather conditions critical to concrete degradation. In the REFI-B project (Jylhä et al. 2011) the FMI also forecast the climates of four localities (coastal area, southern Finland, inland, Lapland) for three periods (until 2030, 2050 and 2100). The forecasts are based on an average of 19 different models which are all based on greenhouse gas emission scenario A2. The A2 scenario involves a situation where greenhouse gases are assumed to increase significantly – it is a sort of worst-case scenario. The FMI also has other significant greenhouse gas emission scenarios: A1B (quite large emissions) and B1 (small emissions).

## 2.3 Number of Annual Freeze-Thaw Cycles

The number of annual freeze-thaw cycles after raining or wet snowing at the most 2 days before freezing in Finnish outdoor climate has been studied in four different weather stations: Helsinki-Vantaa (south coastal area), Jokioinen (south inland), Jyväskylä (inland) and Sodankylä (Lapland). The numbers are calculated with different minimum temperatures to see the difference (Jylhä et al. 2011). The numbers are presented in Table 1, as well as estimated numbers of freeze-thaw cycles in the future climate.

**Table 1** Number of annual freeze-thaw cycles at four different observation stations at most 2 days after rain or sleet events (Jylhä et al. 2011)

| | Temperature under (at most 2 days after rain or sleet) | | | | | |
|---|---|---|---|---|---|---|
| Year and place | 0 °C | −5 °C | −10 °C | 0 °C | −5 °C | −10 °C |
| | Vantaa (south coastal area) | | | Jokioinen (south inland) | | |
| 2000 | 37.8 | 11.7 | 4.0 | 34.6 | 11.1 | 4.2 |
| 2030 | 25.9 | 7.7 | 2.3 | 26.5 | 8.2 | 3.0 |
| 2050 | 21.4 | 6.1 | 1.8 | 23.8 | 7.6 | 2.5 |
| 2100 | 14.5 | 3.9 | 0.4 | 17.2 | 5.8 | 1.1 |
| | Jyväskylä (inland) | | | Sodankylä (Lapland) | | |
| 2000 | 30.4 | 10.4 | 4.2 | 23.4 | 10.4 | 5.0 |
| 2030 | 25.4 | 9.6 | 3.3 | 20.6 | 9.9 | 4.7 |
| 2050 | 24.8 | 9.4 | 3.2 | 22.3 | 11.4 | 5.8 |
| 2100 | 19.8 | 7.2 | 2.1 | 25.2 | 13.3 | 5.7 |

**Table 2** Average change [%] in precipitation compared to present climate (2000) in four different observation station

| | Vantaa (south coastal area | | | Jokioinen (south inland) | | | Jyväskylä (inland) | | | Sodankylä (Lapland) | | |
|---|---|---|---|---|---|---|---|---|---|---|---|---|
| | 2030 | 2050 | 2100 | 2030 | 2050 | 2100 | 2030 | 2050 | 2010 | 2030 | 2050 | 2100 |
| Whole year | 3.8 | 7.1 | 17.7 | 3.1 | 6.1 | 15.2 | 3.6 | 7.1 | 18.2 | 3.9 | 7.9 | 21.3 |

Based on the research made by FMI, the annual rainfall will increase according to Table 2. The change in the amount of rainfall will be higher during autumn and winter when drying of structures is slower in general. For instance, in November the amount of rainfall is going to increase on selected years 2030, 2050 and 2100 in coastal areas by 7.1, 10.9 and 24.4 %, respectively, and on inland 7.9, 10.6 and 27.6 %.

Again, according to FMI, the prevailing wind directions during rain events will stay the same as present. It is concluded that facades faced South-East to West will get more rainfall also in the future.

# 3 Results and Discussion

## 3.1 *Frost Resistance of Concrete*

The success of air entrainment of concrete used in facades and balconies was studied from the samples in the database taken on buildings built 1990 or afterwards because the demands for the concrete grade (C30/37) has been in current level since 1990 (By32 1989). In the database the information related to used air entrainment and its success has been reported as a protective pore ratio ($p_r$), which was common practice in Finland until the year 2004. According to concrete codes (1980), the protective pore ratio should be at least 0.20, which implies that at least 20 % of total porosity of concrete never will be filled by capillary water. A protective pore ratio 0.20 corresponds to a spacing factor 0.25 mm (Koskiahde 2004). Also, in practice for frost damage the concrete needs to freeze to temperatures of $-5\,^{\circ}\mathrm{C}$ or lower (Pigeon and Pleau 1995).

As can be found from Figs. 4 and 5, the air entrainment in fresh concrete has been made with varying success both in facades and balconies. The demand for frost resistance is only fulfilled in approximately 50 % of made panels. If the protective pore ratio is less than 0.1, concrete is not frost resistant in moist conditions, i.e. in ordinary Finnish outdoor during winter time.

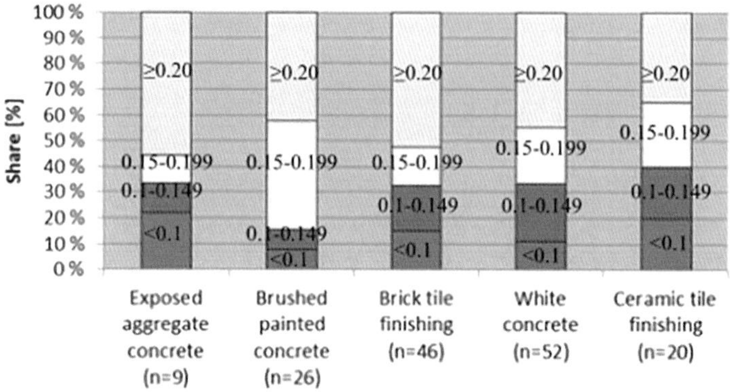


**Fig. 4** Distribution of protective pore ratios in different facade types according to database. The facades are made from 1990 to 1996. Total number of samples is 153

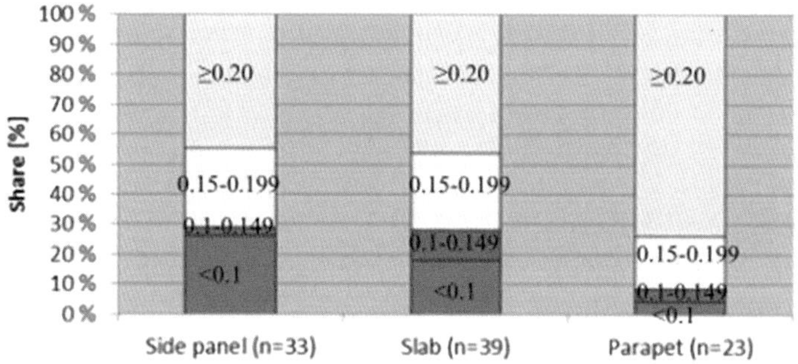


**Fig. 5** Distribution of protective pore ratios in different balcony elements according to database. The balconies are made from 1990 to 1996. Total number of samples is 95

## 3.2 *Frost Damage of Concrete*

The number of freeze-thaw cycles of existing concrete facades and balconies has been studied using following criteria (Lahdensivu 2012):

- raining or wet snowing at most 2 days before freezing
- the number of freeze-thaw cycles when the temperature goes under −5 °C
- studying the existence and state of frost damage with thin-section analyses (245 samples).

Concrete can be very durable in Nordic climate if it has been successfully air-entrained. When concrete was frost resistant ($p_r \geq 0.20$) there were no signs of frost damage in the concrete samples according to thin-section analyses after over 500 freeze-thaw cycles in real outdoor environment. On the opposite, if the protective pore ratio measured from hardened concrete is less than 0.10, the air-entrainment of concrete has failed. Even insufficiently air-entrained concrete

**Table 3** Time to incipient frost damage to reveal in thin-section analyses in different temperature criteria when rain or sleet has come ≤2 days before freezing (pr ≤ 0.10)

| | South coastal area [years] | | Inland [years] | |
|---|---|---|---|---|
| Building year | t ≤ −5 °C | t ≤ −10 °C | t ≤ −5 °C | t ≤ −10 °C |
| 2000 | 26 | 35 | 37 | 45 |
| 2030 | 40 | 61 | 40 | 58 |
| 2050 | 50 | 78 | 41 | 59 |
| 2100 | 79 | 350 | 53 | 90 |

has a service life of some years. Cracking indicating frost damage is the result of an average of 388 freeze-thaw cycles (t ≤ −5 °C) inland (Jyväskylä) and 307 cycles in the southern coastal area (Vantaa). On the southern coastal area this would corresponds to about 22 years and inland into about 24 years (Lahdensivu 2012).

Table 3 shows the estimated time needed for incipient frost damage revealed with thin-section analyses from inadequate frost resistant concrete both inland (Jyväskylä) and south coastal area (Vantaa) based on forecast of the future and above mentioned research. In southern Finland as a consequence of climate change outdoor circumstances, when concrete freezes wet, will ease remarkably already by 2030. In inland the outdoor climate will remain at the present level and the conditions will get even harder with increasing amount of rain and sleet almost to the end of century. The complete failure with air entraining fresh concrete ($p_r \leq 0.10$) will surely lead to frost damage of concrete structure before the chosen service life of the structure (usually at least 50 years).

## 4 Conclusions

The number of freeze-thaw cycles is slightly higher in inland than in south coastal area but the higher amount of rain and sleet in south coastal Finland makes the number of the freeze-thaw cycles needed for the same degree of frost damage much lower than in inland. As a consequence of climate change outdoor circumstances when wet concrete freezes will ease remarkably already by 2030 in southern Finland because the amount of freeze-thaw cycles is going to decrease significantly. In inland the outdoor climate will remain at the present level and the exposure conditions will get even harder with increasing amount of rain and sleet almost to the end of the century. The complete failure with air entraining fresh concrete ($p_r \leq 0.10$) will surely lead to frost damage of concrete structure before the adopted service life of the structure (usually at least 50 years).

The methods used in this study are very simple and do not take into account, for instance, the huge scattering of the data used in determining the occurrence time of incipient and general frost damage. There is also neither consideration on the amount of water which can be absorbed in concrete during rain events nor on how much runs down the surface. Further actions have already started to include the mentioned aspects to the study.

## References

BES (1969) Development of open concrete element system, research report. Suomen Betoniteollisuuden Keskusjärjestö ry (in Finnish)

BY32 (1989) Guidelines for durability and service life of concrete structures. Concrete Association of Finland, Helsinki, 60 p (in Finnish)

Fagerlund G (1977) The critical degree of saturation method of assessing the freeze/thaw resistance of concrete. Tentative RILEM recommendation. Prepared on behalf of RILEM Committee 4 CDC. Materiaux et Constructions, no 58, pp 217–229

IPCC (2007) Climate change 2007: The physical science basis. Contribution of Working Group I to the fourth assessment report of the intergovernmental panel on climate change. Cambridge University Press, Cambridge, UK, 996 pp

Jylhä K, Ruosteenoja K, Räisänen J, Venäläinen A, Tuomenvirta H, Ruokolainen L, Saku S, Seitola T (2009) The changing climate in Finland: estimates for adaption studies. ACCLIM project report 2009, Finnish Meteorological Institute, reports 2009:4, Helsinki 78, p 36 app (in Finnish)

Jylhä K, Ruosteenoja K, Tietäväinen H et al (2011) Rakennusfysiikan ilmastollisten testivuosien sääaineistot nykyisessä ilmastossa ja arviot tulevaisuuden muutoksista. Väliraportti, Finnish Meteorological Institute, Helsinki 6, p 20 app (in Finnish)

Koskiahde A (2004) An experimental petrographic classification scheme for the condition assessment of concrete in facade panels and balconies. Mater Charact 53:327–334

Lahdensivu J (2012) Durability properties and actual deterioration of Finnish facades and balconies, Publication 1028. Tampere University of Technology, Faculty of Built Environment, Tampere, 117 p. 37 app

Lahdensivu J, Mäkelä H, Pirinen P (2013) Durability properties and deterioration of concrete balconies of inadequate frost resistance. Int J Sustain Build Technol Urban Dev 4(2):160–169

Neville A (1995) Properties of concrete. Longman Group, Essex, 844 p

Pentti M (1994) Repair of building envelope. In: Kaivonen J-A (ed) Repair techniques and economy of buildings. Rakennustieto Oy, Saarijärvi, pp 287–358 (in Finnish)

Pentti M, Mattila J, Wahlman J (1998) Repair of concrete facades and balconies. Part 1: structures, degradation and condition investigation, Structural engineering, Publication 87. Tampere University of Technology, Tampere, 156 p (in Finnish)

Pigeon M, Pleau R (1995) Durability of concrete in cold climates. Suffolk. E & FN Spon, London, 244 p

RIL-Finnish Association of Civil Engineers (2013) ROTI-The state of the built environment 2013. RIL, Helsinki, 52 p (in Finnish)

# Influence of Environmental Conditions on Pore Structure Change in Mortar with Various Types of Cement

**Tohru Nakamura, Yukio Hama, and Mohamed Zakaria**

**Abstract** The main aim of this study is to investigate the variation of pore structure and frost resistance in mortar subjected to different environmental conditions and for different exposure times. Mortar specimens were prepared by using three types of cement, which are ordinary Portland cement, fly ash cement, and ground granulated blast furnace cement. Various environmental conditions which are drying, wetting, and drying/wetting cycles were applied to specimens in laboratory, after casting mortar specimens followed by initial curing for 4 weeks in water at 20°. The pore structure modification was evaluated by mercury intrusion porosimetry (MIP) method. It was clarified that the pore structure change can be affected by the studied factors, which are exposure drying temperature, exposure drying duration, and cement type. Furthermore, equations which can predict the pore structure changes was derived based on the experimental results.

**Keywords** Pore structure change • Frost resistance • Drying condition • Cement type

## 1 Introduction

It is well known that the durability of reinforced concrete structures may be decreased due to various environmental conditions such as freezing/thawing or drying/wetting cycles. Concrete structures in cold areas are subjected to frost damage, which can occur due to freezing/thawing cycles in outside environment. High strength concrete without Air entraining (AE) agent was used to achieve a better freezing/thawing durability due to low water-to-cement ratio and high cement content. However, Hama et al. (2002) found that the frost resistance of

T. Nakamura (✉) • Y. Hama
Division of Civil Engineering and Architecture, Muroran Institute of Technology, Muroran, Japan
e-mail: S1921069@mmm.muroran-it.ac.jp

M. Zakaria
Division of Civil Engineering and Architecture, Muroran Institute of Technology, Muroran, Japan

Faculty of Engineering, Aswan University, 81542 Aswan, Egypt

 
C. Andrade et al. (eds.), *Durability of Reinforced Concrete from Composition to Protection*, DOI 10.1007/978-3-319-09921-7_14

high strength concrete without AE excessively decreased after several years of outdoor exposure conditions (high drying temperature and high relative humidity). According to previous studies, Kamada et al. (1986) reported that, by measurements using MIP and ASTM C 666-A testing methods, the volume of pore size 40–2,000 nm is related to durability in hardened ordinary Portland cement paste, with the observation of frost resistance decrease with increasing this pore volume. In the context, Aono et al. (2007) found that, by applying MIP, water vapor absorption and Nuclear Magnetic Resonance (NMR) methods, a change of C-S-H gel nanostructure with 50 °C evaluating temperature, may lead to polymerization of silicate anion chains and development of cohesive structure in C-S-H gel. Shibuya et al. (2010) studied the effect of drying on pore structure change under different conditions of exposure temperature and relative humidity. The results showed that change of pore structure can be strongly affected by drying temperature and drying time than by humidity. Thus, with regard to the change of pore structure, there are some study examples. However, those examples deal with the discussion about the influence of temperature and relative humidity on pore structure variation, but they do not take into account the influence of drying temperature and drying time on pore structure change. Moreover, there are previous studies discussing the influence of various environmental conditions on pore structure change in mortar specimens with ordinary Portland cement; however, there are few studies on the effect of fly ash cement and blast furnace slag cement on pore structure change in mortar.

The aim of this study is to investigate the influence of exposure conditions, which are drying, wetting, and drying/wetting cycles temperatures, and their duration on the change of pore structure in mortar specimens made with three types of cement, which are ordinary Portland cement (OPC), fly ash (FA) cement and blast furnace slag (BFS) cement. Exposure or environmental conditions were applied to mortar in laboratory after initial curing of mortar specimens in water at 20 °C for 4 weeks was completed. Experimental results clarify how the exposure conditions, typically the temperature and duration, and the cement type can influence the change of pore structure in mortar specimens. The obtained findings in the current study are useful for durability issue of cementitious-based materials, since there is a direct relationship between pore structure change and durability evaluation. Finally, equations predicting frost resistance based on pore structure change in mortar with various types of cement are given.

## 2 Experimental Program

Mortar specimens were prepared with the same mix proportions, but cement type. Change in pore structure was investigated before and after subjecting mortar samples, with different cement types, to various exposure conditions in the laboratory. Based on the findings of pore structure change, prediction equations of frost resistance were proposed considering the equations derived by Kamada et al. (1986). More details about the experimental plan are given in this section.

**Table 1** Mix proportions of mortar samples

| Mixture designation | W/B (%) | Mass ratio (wt.%) | | | | | Air content (%) | Flow (mm) |
|---|---|---|---|---|---|---|---|---|
| | | water | Cement (Portland) | Sand | FA | BFS | | |
| N | 35 | 35 | 100 | 200 | – | – | 2.5 | 134 |
| FB | | | 85 | | – | 15 | 1.9 | 139 |
| BB | | | 55 | | 45 | – | 2.1 | 117 |

## 2.1 *Materials and Specimens*

Table 1 shows the mix proportions used in preparation of mortar specimens, which have the mix proportions for sand, water, binder, but cement type. Mortar specimens without AE were casted using three types of cement, which are ordinary Portland cement (N), fly ash cement (FB, 15 % fly ash) and blast furnace slag cement (BB, 45 % slag). In case of FB and BB mortars, 15 and 45 % of cement was replaced by fly ash and furnace slag, respectively, by weight of cement. Land sand from Shiraoi (Hokkaido, Japan) was used as a fine aggregate in mortar with 2.67 g/cm$^3$ dried surface density and 1.57 % for water absorption percentage. Water-binder ratio ($W/B$) and cement-sand ratio were implemented in the mixture as 0.35 and 0.5, respectively. Mortar specimens were with the same dimensions that are W40 × L160 × H40 mm prism size. All mortar specimens were initially cured in water for 4 weeks at 20° after casting, to have the same initial curing condition, before applying drying/wetting conditions.

## 2.2 *Drying/Wetting Conditions*

After 4 weeks of initial curing, mortar samples, mortar specimens were cut into 40 mm cubes, then, subjected to three different environmental or exposure conditions (drying, wetting and drying/wetting cycles) in the laboratory for various exposure temperature and exposure duration. Drying exposure and wetting exposure conditions were applied to mortars at 20, 35 and 50 °C, while drying/wetting cyclinc exposure effect on cement mortars was investigated in a sequence of repeating cycles of drying at 50 °C for 3 days, followed by wetting condition at 50 °C for half a day in water. Pore structure in cement mortars with various types of cement was investigated at different exposure durations or times at 2, 4, 8, 13 and 26 weeks after initial curing of each specimen and for each exposure condition Table 2.

## 2.3 *Experimental Method*

The pore structure of mortar specimens was measured by mercury intrusion porosimetry (MIP) based on JIS R 1655. First, the mortar specimens were cut

**Table 2** Accelerated drying/wetting conditions

| Mixture designation | Initial curing | Drying/wetting condition | Duration (weeks) |
|---|---|---|---|
| 4WK | Water curing at 20 °C for 4 weeks | – | – |
| 20W | | Water at 20 °C | 2, 4, 8, 13 and 26 |
| 20D | | Drying at 20 °C (RH60 %) | |
| 35W | | Water at 35 °C | |
| 35D | | Drying at 35 °C (RH13 %) | |
| 50W | | Water at 50 °C | |
| 50D | | Drying at 50 °C (RH5 %) | |
| 50DW | | Drying/wetting at 50 °C (drying at 50 °C (RH5 %) for 3 days ↔ water at 50 °C for half a day) | |

into 5 mm cubes, soaked in acetone to stop the hydration, and then kept in D-drying condition. Approximately 228 MPa maximum mercury pressure corresponding to 6 nm of minimum pore diameter was applied to 5 mm mortar cubes to measure the pore size distribution in investigated mortar specimens to clarify its relation to exposure temperature and duration.

# 3 Results and Discussion

## *3.1 Pore Size Distribution*

Figure 1 shows the pore size distribution for investigated mortar specimens (N, FB and BB) with various cement type for the conditions of drying, wetting, and drying/wetting cycles exposure at temperatures of 20 and 50 °C and for various exposure duration starting from 0 week until 26 weeks after initial curing has been completed (4 weeks in water at 20 °C). It can be seen from the figure that there is insignificant influence of cement type on the pore structure evolution among the investigated mortar samples in the case of wetting condition and drying conditions. It can be revealed from the pore size distribution results in Fig. 1 that for the case of drying condition at a temperature of 50 °C, the critical pore diameter (at which the peak volume of typical pore size) shifts towards the larger pore diameter region (approximately 100 nm), implying that a high drying temperature has a negative effect on the pore structure, resulting in the coarsening of pore structure, which is in agreement with previous research findings (Atarashi et al. 2009). The trend of shifting of critical pore size can be observed by comparing the critical pore size for the same mortar sample with same conditions, except the exposure time (fo example in 0, 4, 8, 26 weeks), which is important for durability. But, In water condition at 20 and 50 °C, the critical pore diameter shifts towards the smaller pore diameter region, which is in agreement with previous research findings, implying

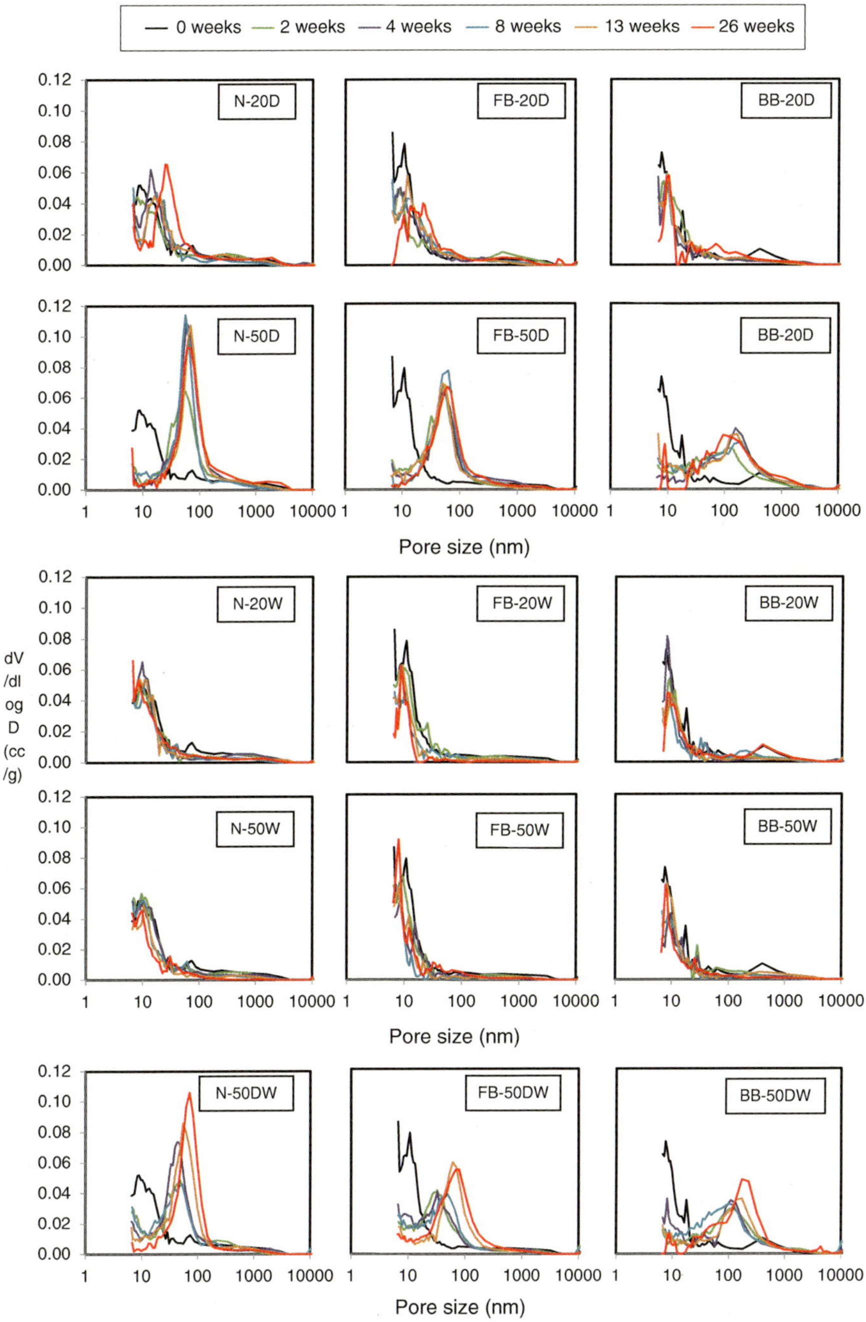


**Fig. 1** Pore size distribution in mortar samples (N, FB and BB) for different drying and wetting exposure temperatures (20, 50 °C) and exposure duration until 26 weeks

that the continuous hydration was occurred. Moreover, in drying/wetting condition (repeat drying at 50 °C for 3 days and water at 50 °C for half a day), it is found that a critical pore diameter more slowly shifts towards the larger pore diameter region (approximately 100 nm) compared to drying case of mortar samples at 50 °C.

## 3.2 Relationship Between Pore Structure Change and Drying Condition

The volume of 40–2,000 nm diameter pores (influential pore size on durability of concrete structures) using MIP has been studied by Shibuya et al. (2010) by applying drying and drying/wetting condition for duration of 2 and 4 weeks. They found that drying exposure can lead to pore-structure coarsening, and that water-supply or wetting exposure condition can cause densification of pore structure within the concrete structure. Thus, both coarsening of pores by drying condition and densification of pores by hydration due to wetting condition are needed to be considered in deep in the current study and for different exposure conditions (drying, wetting, drying/wetting cycles), temperatures (20 and 50 °C) and long duration until 26 weeks.

Figure 2 shows the concept of pore structure change for different exposure conditions such as drying, wetting and drying/wetting cycles conditions (repeating cycles of drying at 50 °C for 3 days and then followed by wetting in water at 50 °C for half a day), which confirm the basis for the current study. Also, Fig. 3 presents change of 40–2,000 nm diameter micro pore volume Influential pore size on durability) in mortar samples with various cement types for different exposure temperatures at 20, 35 and 50 °C until 26 weeks after initial curing (in water at 20 °C for 4 weeks) was finished. Regardless of the type of cement, it can be observed from the figures that under wetting conditions (20, 35 and 50 °C for 26 weeks) the pore structure showed densification, due to the continuous hydration process of cementitious materials in mortar samples, and there was no remarkable

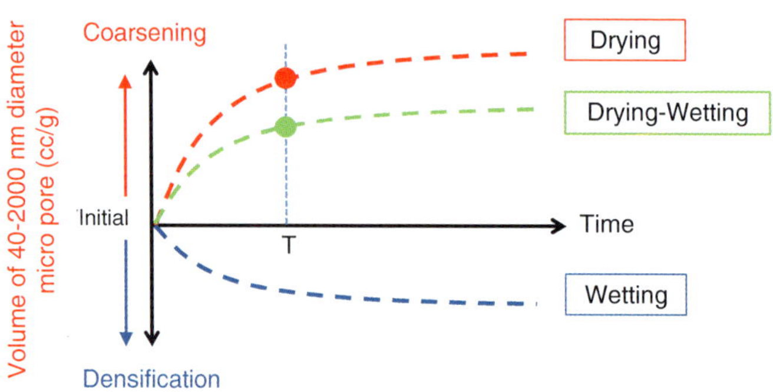


**Fig. 2** Concept of pore structure change for different environmental conditions such as drying, wetting, drying/wetting cycles

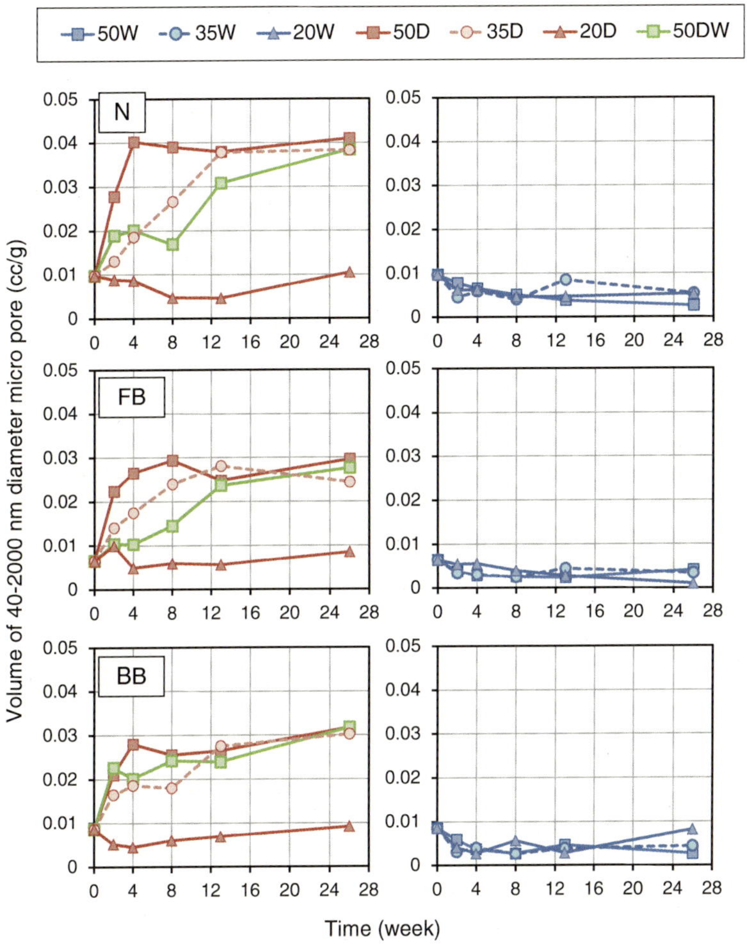


**Fig. 3** Change of 40–2,000 nm diameter micro pore volume with different temperatures at 20 and 50 °C until 26 weeks after initial curing (in water at 20 °C for 4 weeks)

difference of pore structure change for the three cement types in this case. On the other hand, for drying conditions, the volume of 40–2,000 nm diameter micro pore did not change at 20 °C drying case; however, it was increased about three times to four times at 35 and 50 °C drying cases, respectively, compared to 20 °C drying case. Also, for all cement types mortar specimens, it can be seen from Fig. 3 that an upper limit (from 0.03 to 0.04 cc/g) for the change in influential pore structure at 50 °C drying case around 4 to 8 weeks drying exposure duration. Also, it is clear from the figure that when mortar specimens were subjected to drying/wetting condition (repeating cycles of drying at 50 °C for 3 days and then followed by wetting at 50 °C for half a day), the volume of 40–2,000 nm diameter pores increased to the upper limit for the case of pore structure change by drying condition at 50 °C. Thus, the concept of influential pore structure trend shown in Fig. 2 can be confirmed from the test results in this study. From the observation, it can be seen that the concept of pore structure change in mortar samples subjected to drying and drying/wetting conditions can be validated by experimental data.

## 4 Deriving the Prediction Equation of Pore Structure Change

Aono et al. (2007) and Shibuya et al. (2010) mentioned that pore structure change can be affected by cement type, temperature, exposure duration and humidity. The maturity method has been widely used to estimate the temperature dependency of strength development in concrete practically. It considers that the pore structure change (coarsening or densification of pores) is temperature dependent; therefore, it is possible that the influence of temperature and age on the pore structure change is related to the Maturity.

Mc Intosh (1949) reported that the temperature dependency of strength development in concrete can be described by the concept of maturity, defined as in Eq. 1 using datum temperature.

From the obtained results (Fig. 3), it was found that there is temperature dependency for the change of pore structure (the volume of 40–2,000 nm diameter micro pore) in drying condition. Thus, it is considered that, it can be regarded that there is a datum temperature for pore structure development and the temperature dependency has the same form as in Eq. 1.

$$M = \sum^{n} (\theta - DT) \tag{1}$$

where, $M$ is maturity (°D · D), $\theta$ is the temperature (°C), $DT$ is the datum temperature (°C), $n$ is the day (day).

Moreover, as shown from the results in Fig. 3, it is clear that there is an upper limit to the amount of pore structure change by drying condition. Here, in this study, the characteristic of pore structure change ($PVd/PVi$) is expressed as the ratio of the volume of 40–2,000 nm diameter micro pore in drying condition, $PVd$, to its volume after initial curing condition, $PVi$, (in water at 20 °C for 4 weeks). Therefore, the relationship between the pore structure change ($PVd/PVi$) and the maturity ($M$) can be simulated as shown in Fig. 4.

Firstly, the maximum value of pore structure change (upper limit) was calculated from values obtained from pore structure change due to drying at 50 °C for 8, 13, 26 weeks, and at 35 °C for 26 weeks for all mortar samples. Maximum values of the change of pore structure ($PVd/PVi$) for mortar specimens are given in Table 3. Secondly, the relationship between the pore structure change ($PVd/PVi$) and the maturity ($M$) can be shown by a linear regression. Then, the datum temperature ($DTd$) in Eq. 1, at which the pore structure changes, was assumed in the range of

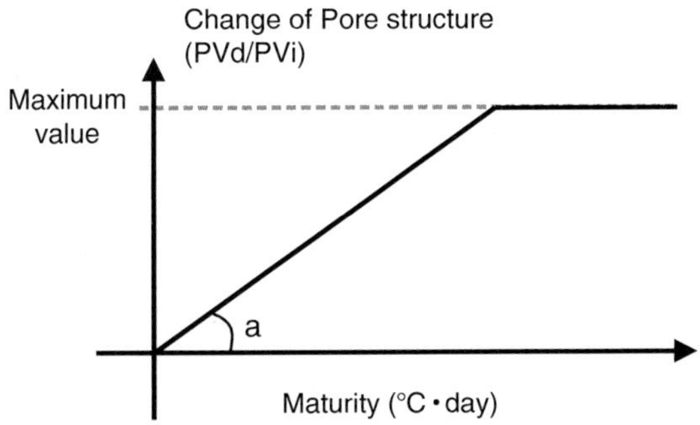


**Fig. 4** Model for the relationship between pore structure change and maturity due to drying

**Table 3** Maximum value of pore structure change (PVd/PVi)

| Cement type | Maximum value of pore structure change (PVd/PVi) |
|---|---|
| N | 4.026 |
| FB | 4.162 |
| BB | 3.308 |

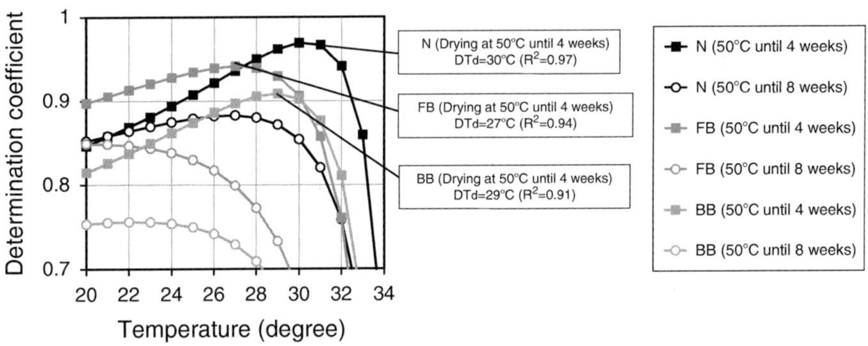


**Fig. 5** Change of determination coefficient for the linear regression

temperature from 20 °C to 34 °C. After that, a slope and determination coefficient of the regression were calculated from these results for the change of the volume of 40–2,000 nm diameter micro pore in drying at 35 °C by 13 weeks and at 50 °C by 4 or 8 weeks. The correlation was good in the case of the square root of temperature dependency ($\sqrt{Mp}$) and logarithm natural of pore structure change ($ln(PV_d/PV_i)$).

Figure 5 shows the change of determination coefficient for the linear regression. For N samples, in the case of drying condition at 50 °C for 4 weeks, the datum temperature ($DT_d$) was 30 °C and the determination coefficient was 0.97. For FB samples, in the case of drying condition at 50 °C for 4 weeks, the datum temperature ($DT_d$) was 27 °C and the determination coefficient was 0.94. Moreover, for BB samples, with the same drying conditions, the datum temperature ($DT_d$) was 29 °C

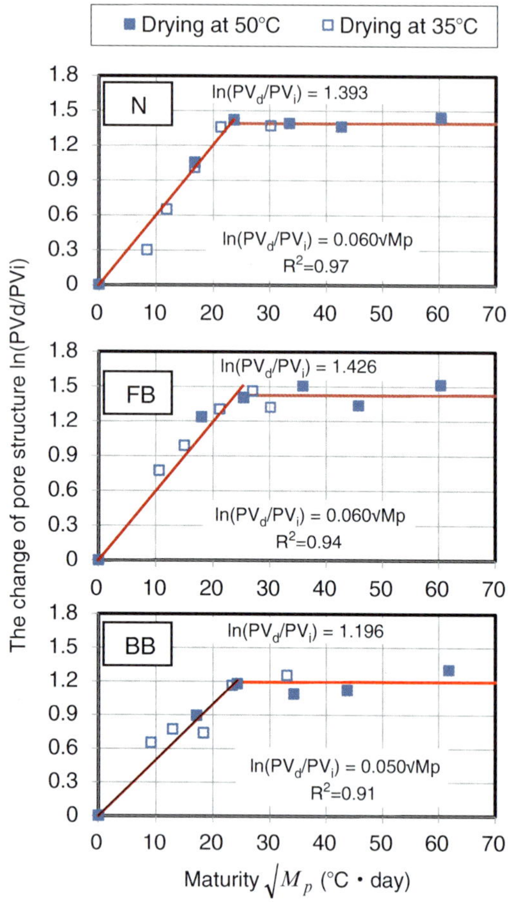


**Fig. 6** Relationship between the temperature dependency and pore structure change and the experimental date under the terms of the highest determination coefficient

and the determination coefficient was 0.91. Thus, for all cement types, the datum temperature ($DT_d$) was almost the same and a noticeable difference of the tendency of pore structure change was not found.

Figure 6 shows the relationship between the pore structure change (*PVd*/*PVi*) and the maturity ($M$) and the experimental date in the highest determination coefficient. For all cement types, it was found that the line of regression corresponds well to the experimental data.

$$ln\left(\frac{PV_d}{PV_i}\right) = a\sqrt{Mp} \tag{2}$$

where, *PVd*/*PVi* is the pore structure change in drying condition, *PVd* is the volume of 40–2,000 nm diameter micro pore in drying condition, *PVi* is the volume of 40–2,000 nm diameter micro pore in initial curing, *a* is constant value (which for ordinary Portland cement (N) is 0.060, for fly ash cement (FB) is 0.060 and for blast furnace slag cement (BB) is 0.050), *Mp* is the maturity contributing to the pore structure change by drying (°C · day) as defined in Eq. 1.

In this paper, the proposed equation is limited to the condition (*W*/*B* is 0.35), and it does not describe the influence of the ratio of water to binder on the pore structure change. Nakamura et al. (2013) reported that the pore structure change (*PVd*/*PVi*) can be obtained as a function of the ratio of water to binder. Thus, it was assumed that the constant a is a function of the ratio of water to binder. Hence, further accumulation of data is necessary for verification of equation.

Figure 7 shows the correlation of the predicting equations to the experimental data in the drying/wetting condition (repeat drying at 50 °C for 3 days and water at 50 °C for half a day). It is clear from the figure that the upper limit of the amount of pore structure change is same as in the drying condition, however, there is gradual change in pore structure. Thus, it is considered that it is necessary to correct for the influence of water condition and humidity on the pore structure.

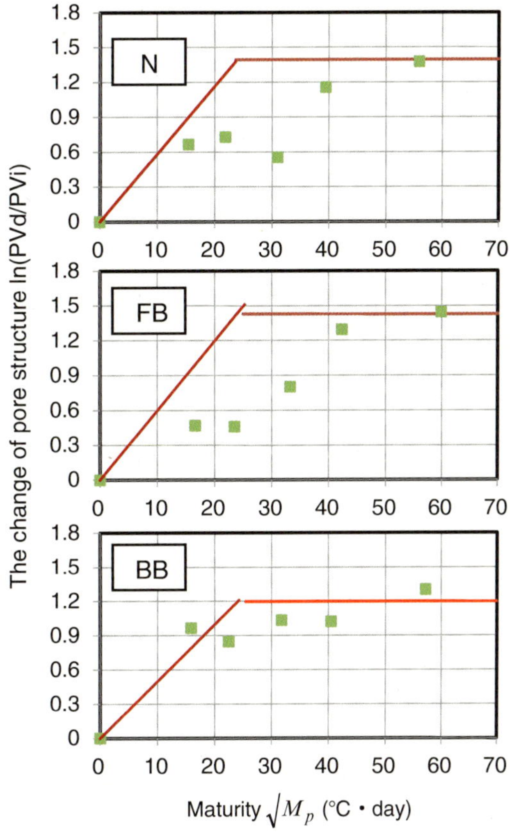


**Fig. 7** The correlation of the predicting equations to the experimental data in drying/wetting condition (repeat drying at 50 °C for 3 days and water at 50 °C for half a day).

Also, Kamada et al. (1986) explained that the frost resistance can be affected by the volume of 40–2,000 nm diameter pores, for that Eq. 3 was given, which presents a durability factor as a function of the volume of 40–2,000 nm diameter pores and air content

$$logDF = -0.317 - 1.209 logPV + 1.799A/P \tag{3}$$

where, *DF* is the durability factor, *PV* is the volume of 40–2,000 nm pore diameter (cc/g), *A*/*P* is air content to past ratio (%).

Then, the relationship of the change of durability factor ($DF_d/DF_i$) and pore structure change due to drying can be described by Eq. 4.

$$DF_d/DF_i = \left(PV_d/PV_i\right)^{-1.209} \tag{4}$$

where, $DF_d$ is the durability factor after drying exposure, $DF_i$ is durability factor at initial condition (in water for 4 weeks at 20 °C) and before drying exposure.

Moreover, the change of durability factor ($DF_d/DF_i$) can be predicted as in Eq. 5 by considering drying temperature and drying duration factors.

$$DF_d = DF_i\left\{exp\left(a\sqrt{M}\right)\right\}^{-1.209} \tag{5}$$

## 5 Conclusions

1. Regardless of cement type, the pore structure reached densification due to hydration in the case of wetting conditions until 26 weeks at either 20, 35 and 50 °C.
2. The volume of 40–2,000 nm diameter micro pores increased to the upper limit of pore volume change by the drying condition at both 35 and 50 °C, implying that upper limit of coarsening can be reached in a relatively high temperature.
3. When the mortar specimens were exposed to drying exposure condition at 20 °C, change of the volume of 40–2,000 nm diameter micro pores was insignificant. However, a significantly large pore volume, which is three to four times of 20 °C drying condition volume of 40–2,000 nm diameter micro pores, was observed for the case of drying exposure condition at both 35 and 50 °C.
4. When the mortar specimens were exposed to drying/wetting condition (repeating cycles of drying at 50 °C for 3 days, followed by wetting by keeping in water at 50 °C for half a day), the volume of 40–2,000 nm diameter micro pores was also increased to the upper limit of the amount of pore volume change by the drying condition, however, there is gradual change in pore structure, and it is necessary to correct for the influence of water condition and humidity on the pore structure change.

5. Predicting equations which can estimate the change of pore structure in mortars with different cement type were derived based on the experimental results of 40–2,000 nm diameter micro pores diameter in drying exposure condition at 35 and 50 °C. Also, proposed equations can be related to drying temperature and drying duration.
6. Predicting equations of the change of durability factor in consideration of these factors (drying temperature and age) were derived based on the change of 40–2,000 nm diameter micro pores each cement type.

## References


Yoshimichi Aono, Fumiaki Matsushita, Sumio Shibata, Yukio Hama (2007) Nano-structure changes of C-S-H in hardened cement paste during drying at 50 °C. J Adv Concr Technol 5(3):313–323

Daiki Atarashi, Yikio Hama, Masaru Shibuya, Yoshimichi Aono (2009) Changes of pore structure and frost resistance of mortar during drying or wetting-drying. Cem Sci Concr Technol (63):155–160

Yukio Hama, Eisuke Hamada, Osamu Senbu. Tomosawa Fuminori (2002) Effect of curing condition before freezing and thawing test to frost resistance of high strength concrete and high fluidity concrete. Cem Sci Concr Technol (56):425–430

Eiji Kamada, Osamu Senbu, Masayuki Tabata, Hirokazu Tanaka (1986) Statistical investigation concerning the effects of pore structure on the frost resistance of concrete. J Struct Constr Eng AIJ (487):1–9

Mc Intosh JD (1949) Electrical curing of concrete. Mag Concr Res 2(2):79–88

Tohru Nakamura, Yukio Hama, Yoshihiko Kishimoto, Mohamed Zakaria (2013) Proposing prediction method of frost resistance based on the change of pore structure in mortar subjected to drying. Japan Concrete Institute, Tokyo

Masaru Shibuya, Yukio Hama, Daiki Atarashi, Yoshihiko Kishimoto (2010) Relationship between pore structure change and frost resistance of mortar after outdoor exposure, Summaries of technical papers of annual meeting architectural institute of Japan, No 1638, pp 1275–1276

# Study on the Chloride Diffusion Coefficient in Concrete Obtained in Electrically Accelerated Tests

**P. Spiesz and H.J.H. Brouwers**

**Abstract** This study presents an analysis of the chloride diffusion coefficient ($D_{RCM}$), obtained in electrically accelerated chloride migration tests. As demonstrated here, the obtained chloride diffusion coefficient does not represent the apparent one, as it is independent of chloride binding. This is accounted for the fact that the $D_{RCM}$ is calculated based on the average chloride penetration depth measured in concrete and binding at the position of the chloride penetration front is very limited. It is also demonstrated in this study that the apparent chloride diffusion coefficient in concrete is not constant within the sample, because it depends upon the local chloride concentrations.

**Keywords** Diffusion • Chloride • Transport properties • Modeling • Durability

## 1 Introduction

In view of problems in practice with chloride-induced corrosion of reinforcing steel, there is a need for quantified information on chloride transport properties of concrete. Thus, a reliable prediction model for chloride ingress into the concrete cover is considered as the key point for an assessment of the long-term behaviour of concrete exposed to sea water or de-icing salts. A number of laboratory test methods (both long- and short-term) have been developed to quantify the chloride transport in concrete. The long-term methods, however, are often not preferred from a practical point of view, because they are laborious, time consuming and costly. To overcome these disadvantages, an electrically-accelerated test – the Rapid Chloride Migration test (RCM), also known as CTH or RMT, has been developed by Tang (1996) and released as a guideline NT Build 492 (1999). Within

P. Spiesz (✉)
Department of the Built Environment, Eindhoven University of Technology, Eindhoven, The Netherlands

ENCI HeidelbergCement Benelux, The Netherlands
e-mail: p.spiesz@tue.nl; przemek.spiesz@heidelbergcement.com

H.J.H. Brouwers
Faculty of the Built Environment, Eindhoven University of Technology, The Netherlands
e-mail: jos.brouwers@tue.nl



C. Andrade et al. (eds.), *Durability of Reinforced Concrete from Composition to Protection*, DOI 10.1007/978-3-319-09921-7_15

the European DuraCrete report the output value of the RCM test: $D_{RCM}$ – chloride diffusion coefficient obtained in migration tests – has been introduced into the service life model for concrete, which has resulted in even higher popularization of the test method. The application of the $D_{RCM}$ for service life modelling of concrete puts strict requirements on the test such as its theoretical soundness, reliability and repeatability of results. Yet, many researchers concluded that the $D_{RCM}$ should be treated with scepticism, especially when used for the service life design of concrete. One reason for this scepticism is the discrepancy between the theoretical and experimental chloride concentration profiles in concrete, presented in many studies (Hooton et al. 2001, Stanish 2002, Yuan 2009, Spiesz et al. 2012, Spiesz and Brouwers 2012).

Spiesz et al. (2012) attempted to extend the traditional chloride transport model for the RCM test by taking into account the non-linear binding of chloride in concrete and non-equilibrium conditions between free and bound chloride concentrations. As a result, they were able to explain the shape of measured chloride concentration profiles in concrete after the RCM test and obtain, among other parameters, the effective chloride diffusion coefficient in concrete. In the present study a further analysis of the chloride diffusion coefficient obtained in the RCM test is performed. To do so, the experimental data is applied to both, the traditional RCM chloride transport model and the extended model presented in Spiesz et al. (2012). Subsequently, the apparent chloride diffusion coefficient ($D_{app}$), obtained from the extended chloride transport model is compared to the traditional $D_{RCM}$.

## 2 Traditional and Extended Chloride Transport Models

The chloride transport model for the RCM test, developed by Tang (1996) reads as follows:

$$\frac{\partial c}{\partial t} = \frac{D_0}{1 + \frac{\partial c_b}{\partial c}} \left( \frac{\partial^2 c}{\partial x^2} - \frac{zFU}{RTL} \frac{\partial c}{\partial x} \right) = D_{RCM} \left( \frac{\partial^2 c}{\partial x^2} - \frac{zFU}{RTL} \frac{\partial c}{\partial x} \right) \tag{1}$$

where: $c$ – free chloride concentration [g/dm$^3$], $t$ – time [s], $D_0$ – intrinsic chloride diffusion coefficient [m$^2$/s], $c_b$ – bound chloride concentration [g/dm$^3$], $x$ – distance [m], $z$ – ion valence, $F$ – Faraday constant [C/mol], $U$ – voltage [V], $R$ – universal gas constant [J/(mol · K)], $T$ – temperature [K], $L$ – thickness of the sample [m] and $D_{RCM}$ – chloride diffusion coefficient determined in migration tests [m$^2$/s].

A constant value of $\partial c_b/\partial c$ term assumed in Eq. 1 implies that either no binding or instantaneous linear chloride binding (equilibrium conditions) is involved in the chloride migration process. The $D_{RCM}$ implicitly takes binding into account, as shown in Eq. 1, so it represents the apparent diffusion coefficient ($D_a$). A more detailed definition of the $D_a$ will be presented later, in Eq. 7.

Tang (1996) proposed a solution of Eq. 1, eventually yielding the following equations for the calculation of $D_{RCM}$ based on the measured chloride penetration depth:

$$D_{RCM} = \frac{RTL}{zFU} \cdot \frac{x_d - \alpha\sqrt{x_d}}{t} \tag{2}$$

$$\alpha = 2\sqrt{\frac{RTL}{zFU}} \cdot erf^{-1}\left(1 - \frac{2c_d}{c_0}\right) \tag{3}$$

where: $x_d$ – chloride penetration depth indicated by the colourimetric indicator [m], $\alpha$ –laboratory constant, $c_d$ – chloride concentration at which the colourimetric indicator changes the colour [g/dm$^3$] and $c_0$ – chloride concentration in the external electrolyte [g/dm$^3$].

In contrast to the mathematical model given in Eq. 1, the extended chloride transport model for the RCM test developed by Spiesz et al. (2012), accounts for non-linear chloride binding and concentrations non-equilibrium, and reads as follows:

$$\varphi \frac{\partial c}{\partial t} + u \frac{\partial c}{\partial x} = -k \left[ c - \left( \frac{C_b}{K_b} \right)^{\frac{1}{n}} \right] \tag{4}$$

$$(1 - \varphi)\rho_s \frac{\partial C_b}{\partial t} = k \left[ c - \left( \frac{C_b}{K_b} \right)^{\frac{1}{n}} \right] \tag{5}$$

where: $\varphi$ – porosity, $u$ – ionic migration velocity [m/s], $k$ – chloride mass transfer rate [1/s], $\rho_s$ – specific density of concrete [g/dm$^3$], $C_b$ – bound chloride concentration [g/g], $K_b$ – chloride binding capacity [dm$^{3n}$/g$^n$] and $n$ – chloride binding intensity. The abovementioned system of equations considers separately the free (Eq. 4) and bound (Eq. 5) chloride concentrations.

The ionic migration velocity $u$ can be further defined as:

$$u = \frac{D_{eff} zFU}{RTL} \tag{6}$$

where: $D_{eff}$ – effective chloride diffusion coefficient [m$^2$/s].

As can be seen in Eqs. 4, 5, and 6, the effective chloride diffusion coefficient in concrete is used in order to treat independently the binding term and to express the diffusion coefficient in the terms of the flux per unit area of concrete rather than only the pore solution.

## 3 Experimental Data

The RCM test data for five different concrete mixtures: C1 and C2 (Stanish 2002), C4–C6 (Yuan 2009) and one mortar: C3 (Spiesz and Brouwers 2012) were used in this study. The mixture proportions and hardened state properties of concrete are presented in Table 1. The RCM test conditions are given in Table 2.

**Table 1** Mix proportions and properties of the analyzed concretes

| | C1 | C2 | C3 | C4 | C5 | C6 |
|---|---|---|---|---|---|---|
| Mix number | Composition [kg/m$^3$] | | | | | |
| Cement | 418 | 363 | 786 | 400 | 380 | 363 |
| Water | 146 | 163 | 207 | 140 | 182 | 218 |
| Aggregates | 1,025 | 1,025 | – | 1,281 | 1,217 | 1,162 |
| Sand | 695 | 698 | 1,382 | 660 | 627 | 599 |
| $w/c$ | 0.35 | 0.45 | 0.26 | 0.35 | 0.48 | 0.60 |
| $\rho_s$ [g/dm$^3$] | 2,601 | 2,606 | 2,710 | 2,665 | 2661 | 2,643 |
| $\varphi$ [%] | 11.3 | 14.3 | 15.0 | 10.2 | 14.1 | 15.0 |
| Cement type | T10 | T10 | CEM I 52.5 N | CEM I 52.5 N | CEM I 52.5 N | CEM I 52.5 N |

**Table 2** RCM test conditions for the analyzed experiments

| Mix number | $t$ [h] | $U$ [V] | $c_0$ [g/dm$^3$] |
|---|---|---|---|
| C1 a | 6 | 60 | 64.95 |
| C1 b | 9 | 60 | 64.95 |
| C1 c | 18 | 60 | 64.95 |
| C2 a | 6 | 60 | 64.95 |
| C2 b | 9 | 60 | 64.95 |
| C2 c | 18 | 60 | 64.95 |
| C3 a | 24 | 35 | 64.95 |
| C3 b | 24 | 47.5 | 64.95 |
| C3 c | 24 | 60 | 64.95 |
| C4 | 35 | 35 | 64.95 |
| C5 | 25 | 25 | 64.95 |
| C6 | 20 | 20 | 64.95 |

After every performed RCM test (Table 2), also the total chloride concentration profile in the sample was determined. Following the procedure described in Spiesz et al. (2012), using the experimental data presented in Tables 1 and 2 and the measured chloride concentration profiles, the parameters $D_{eff}$, $k$, $K_b$ and $n$ in Eqs. 4 and 5 could be obtained from fitting the computed chloride concentration profiles to the measured ones.

Figure 1 presents the fitted profile for concrete C1a as an example.

Table 3 presents the $D_{RCM}$ values calculated from Eq. 2 as well as the $D_{eff}$ obtained from empirical fitting of Eqs. 4 and 5 to the experimental data.

## 4 Analysis of the Chloride Diffusion Coefficient

The $D_{RCM}$, as defined in Eq. 1, represents the apparent diffusion coefficient ($D_{app}$) and can be further presented as follows (Atkinson and Nickerson 1984, Nilsson et al. 1994):

$$D_{app} = \frac{D_0}{1 + \frac{\partial c_b}{\partial c}} = \frac{D_{eff}}{\varphi\left(1 + \frac{\partial c_b}{\partial c}\right)} \tag{7}$$

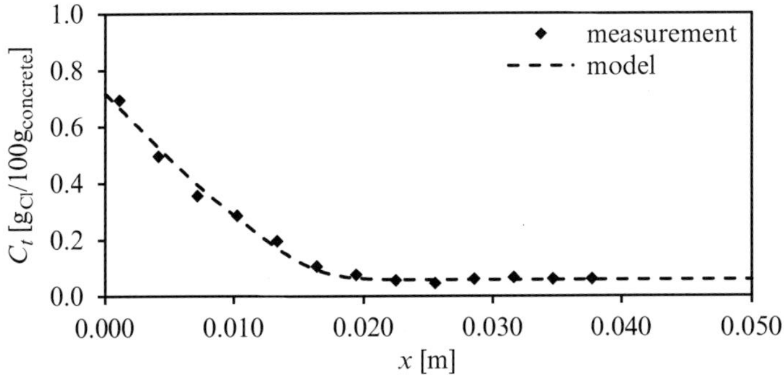


**Fig. 1** Measured and computed total chloride concentration profiles in concrete C2a. $C_t$ represents the total chloride concentration in concrete in wt. % or $g_{Cl}/100g_{concrete}$

**Table 3** $D_{RCM}$ and effective chloride diffusion coefficients ($D_{eff}$)

| | $x_d$ [mm] | $D_{RCM}$ [$\cdot 10^{12}$ m$^2$/s] | $D_{eff}$ [$\cdot 10^{12}$ m$^2$/s] |
|---|---|---|---|
| C1a | 7.85 | 6.85 | 0.81 |
| C1b | 14.70 | 8.91 | 1.05 |
| C1c | 26.85 | 8.24 | 0.97 |
| C2a | 15.99 | 14.60 | 1.94 |
| C2b | 22.61 | 13.99 | 1.91 |
| C2c | 42.19 | 13.35 | 1.80 |
| C3a | 11.05 | 4.16 | 0.75 |
| C3b | 13.52 | 3.84 | 0.67 |
| C3c | 19.10 | 4.40 | 0.79 |
| C4 | 15.70 | 6.08 | 0.65 |
| C5 | 16.88 | 9.16 | 1.21 |
| C6 | 23.54 | 16.45 | 2.22 |

An identical definition of the apparent chloride diffusion coefficient as shown in Eq. 7 has been also adopted in the solution of Fick's 2nd law for the diffusion of chlorides in concrete in long term chloride diffusion tests (e.g. the immersion tests – NT Build 443 1995, ASTM C1556-4 2004). Nevertheless, it has to be mentioned that the $D_a$ computed later in this study for chloride migration tests has not been compared to the apparent chloride diffusion coefficient determined in long term chloride immersion tests.

As explained earlier, the $D_{RCM}$ is assumed to be constant in the basic RCM test chloride transport model (Eq. 1), which implies a constant binding capacity term ($\partial c_b/\partial c$), i.e. no binding or instantaneous linear chloride binding (equilibrium conditions) take place during the chloride migration process. Such an assumption is a simplification, as it is widely accepted that chloride binding in concrete is non-linear and relatively long time (up to 10 days) is needed to complete binding (reach the equilibrium conditions), while the duration of the RCM test in most cases is limited just to 24 h.

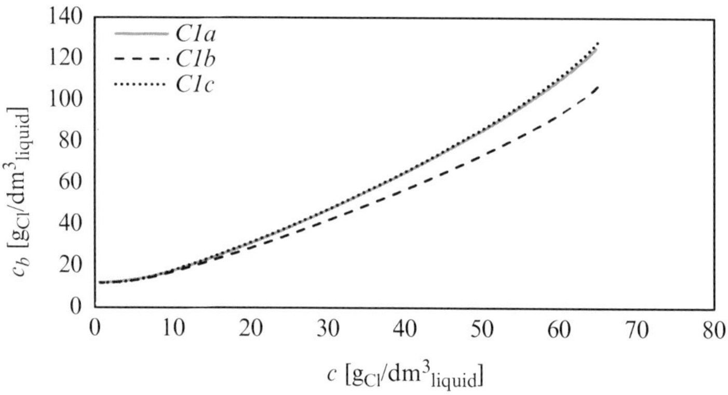


**Fig. 2** Chloride binding curves computed from the model (Eqs. 4 and 5)

In order to determine the apparent chloride diffusion coefficient from the extended chloride transport model, Eq. 7 can be used. Hence, the $D_{eff}$ and free and bound chloride concentration profiles ($c$ and $C_b$, respectively) were obtained by applying the model (Eqs. 4 and 5) to the experimental data (more details can be found in Spiesz et al. 2012). Since the unit of $C_b$ is $g_{Cl}/g_{solid}$ and in the term $\partial c_b/\partial c$, needed for the calculation of the $D_{app}$ (Eq. 7), both concentrations should have consistent units (i.e. $g_{Cl}/dm^3_{liquid}$), the expression $c_b = C_b \cdot \rho_a/\varphi$ was used to unify the units ($\rho_a$ – apparent density of concrete [g/dm$^3$]).

With known $c_b$ and $c$ profiles in concrete, the binding isotherm during the RCM test can be derived. In Fig. 2 the chloride binding curves obtained from the model for concrete C1a-c are shown as an example. It can be noticed from the binding curves that during migration tests binding is very limited at low free chloride concentrations. The reduced binding at low $c$ can be explained through the reaction model given in Eqs. 4 and 5 (Spiesz et al. 2012). In this model, the reaction (binding) rate is proportional to the deviation of the chloride concentration in the bulk solution from the equilibrium concentration at the liquid–solid interface. Therefore, the deviation of $c_b$ from the equilibrium concentration will be low for low values of $c$ and it will increase proportionally with an increase of $c$. In turn, the increase in the amount of bound chlorides will be low at low $c$ and respectively higher for higher $c$.

The term $\partial c_b/\partial c$, needed for the estimation of $D_{app}$ (Eq. 7), was derived from the computed free and bound chloride concentration profiles, and is shown in Fig. 3 for concrete C1a–c. Again, one can see that the binding capacity, defined as $\partial c_b/\partial c$, is greater at higher free chloride concentrations than at lower concentrations, as explained previously.

With known porosities (from measurements) and free and bound chloride concentration profiles as well as the effective diffusion coefficients (obtained from the model), it is possible to estimate the apparent diffusion coefficient profiles in concrete, applying Eq. 7. Figure 4a–b show the computed $D_{app}$ profiles, obtained for concretes C1 and C2, respectively, for which the test conditions were the same,

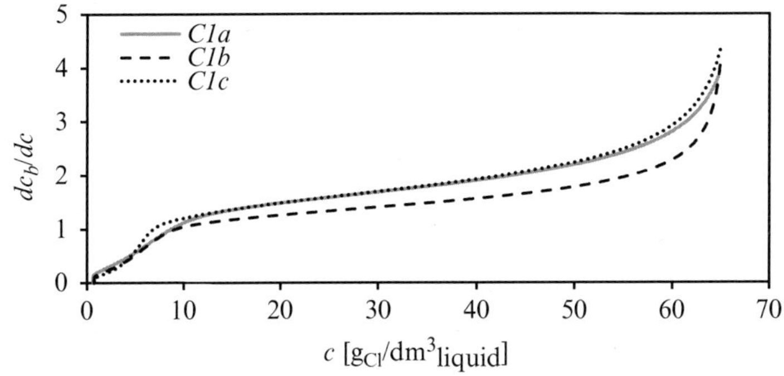


**Fig. 3** Chloride binding capacity versus free chloride concentration

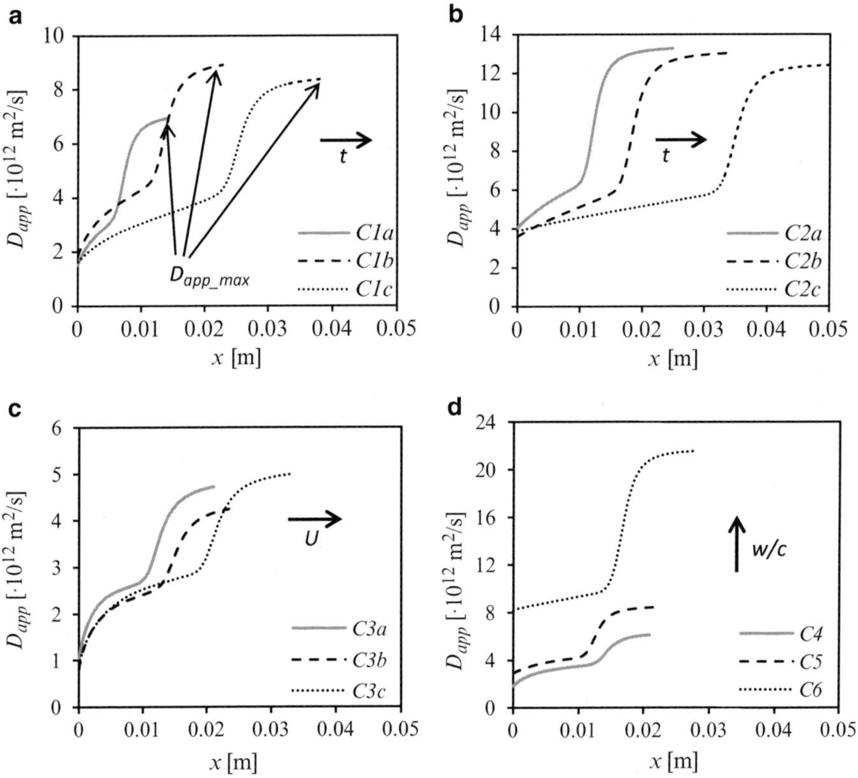


**Fig. 4** The apparent chloride diffusion coefficient profiles during the chloride migration tests; (**a**) C1a–c, (**b**) C2a–c, (**c**) C3a–c and (**d**) C4, C5 and C6

except for the duration (i.e. 6, 9 and 18 h of the RCM test at 60 V). Figure 4c shows the $D_{app}$ profiles obtained for mortars C3a–c, for which the duration of the test was the same for all the samples (24 h), but the applied voltages were different (35, 47.5 and 60 V). Figure 4d shows the $D_{app}$ profiles obtained for concretes with different

$w/c$ ratios (0.35, 0.48 and 0.6). The $D_{app}$ profiles given in Fig. 4a–d are computed until the maximum free chloride penetration depth (the so-called chloride penetration front). All the derived $D_{app}$ profiles have a similar characteristic: lower values of the $D_{app}$ in the layers of concrete closer to the exposed surface (higher chloride concentrations), followed by a region of a rapid increase and finally a plateau in the layers close to the chloride penetration front (very low chloride concentrations). Due to the increased chloride concentrations in the surface layers, the term $\partial c_b/\partial c$ is also larger in these layers (see Fig. 3), which in turn reduces the value of the $D_{app}$, following Eq. 7. The regions of increased $D_{app}$ values in Fig. 4a–d correspond to the value of the binding capacity $\partial c_b/\partial c$, which reduces significantly at low free chloride concentrations, as can also be seen in Fig. 3. For very low chloride concentrations the binding capacity is also low and it does not significantly influence the value of the $D_{app}$, which explains the plateau of the profiles in the vicinity of the chloride penetration front (see Fig. 4a–d). One can notice in Fig. 4a–d that in each profile the $D_{app}$ is not constant as assumed in the basic RCM test model (Eq. 1), but instead varies greatly: the difference between the minimum values (in the regions of concrete with high $c$) and the maximum values (vicinity of the chloride penetration front, very low $c$) is of about a factor of four.

The maximum values of the apparent chloride diffusion coefficient ($D_{app\_max}$), computed from the model for each derived profile shown in Fig. 4a–d, are given in Table 4. These values represent the diffusion coefficient at very low free chloride concentrations, right behind the progressing chloride penetration front in concrete, where the binding capacity is very low. Following Eq. 7 it can be stated with a good approximation that the $D_{app\_max}$ equals the $D_{eff}/\varphi$ or $D_0$, as the values of the term $\partial c_b/\partial c$ are negligible for the progressing chloride penetration front.

As shown in Table 4, the values of the $D_{RCM}$ obtained from the basic RCM model are very similar to the values of the $D_{app\_max}$ retrieved from the computed distribution of the $D_{app}$ in concrete. This means that the $D_{RCM}$ represents the apparent chloride diffusion coefficient in concrete only in the vicinity of the

**Table 4** $D_{RCM}$ versus $D_{app\_max}$ for concretes C1–C6

| | $D_{RCM}$ [$\cdot 10^{12}$ m$^2$/s] | $D_{app\_max}$ [$\cdot 10^{12}$ m$^2$/s] |
|---|---|---|
| C1a | 6.85 | 6.94 |
| C1b | 8.91 | 8.92 |
| C1c | 8.24 | 8.38 |
| C2a | 14.60 | 13.28 |
| C2b | 13.99 | 13.03 |
| C2c | 13.35 | 12.40 |
| C3a | 4.16 | 4.73 |
| C3b | 3.84 | 4.25 |
| C3c | 4.40 | 5.01 |
| C4 | 6.08 | 6.12 |
| C5 | 9.16 | 8.40 |
| C6 | 16.45 | 21.53 |

chloride penetration front (low values of $c$), where the binding of chlorides is very limited. As can be concluded from the binding curves shown in Fig. 3, the chloride penetration front in concrete is only slightly retarded by chloride binding. This influence becomes more significant when the chloride concentration increases, in which case, the value of the $D_{app}$ behind the chloride penetration front decreases towards the exposed surface, where the $D_{app}$ is the lowest (see Fig. 4a–d).

Based on the explanations given above, Eq. 7, which represents the $D_{RCM}$ in the basic RCM model, can be modified for the RCM test by neglecting the chloride binding capacity at the location of the chloride penetration front:

$$D_{RCM} = D_{app\_max} \approx D_0 \approx \frac{D_{eff}}{\varphi} \tag{8}$$

It can be stated that the assumption of a constant $D_{RCM}$ adopted in the basic RCM model is correct; however, not because of the constant binding capacity term assumed in this model, but due to the lack of chloride binding at very low chloride concentrations. One can notice that the $D_{RCM}$ represents only one case of the $D_{app}$, i.e. when there is no chloride binding, and then it equals the $D_{eff}/\varphi$ or $D_0$. Therefore, the constant $D_{RCM}$, as defined in Eq. 1, is not valid in the entire volume of the tested concrete sample, but only in the vicinity of the chloride penetration front, where the binding is negligible. Hence, as the $D_{RCM}$ is determined using the position of the chloride penetration front, it should not be affected by binding.

The correlation between the $D_{RCM}$ and $D_{eff}$ given in Eq. 8 is very straightforward – only the porosity is the proportionality factor between these two coefficients. This relationship is presented in Fig. 5 for the data analyzed in this study. The linear correlation between the $D_{RCM}$ and $D_{eff}$ shown in Fig. 5 is clear and the proportionality factor of 0.133 obtained in the regression can also be predicted, as it is the weighted average porosity of all the concrete samples analyzed in this study (i.e. 0.134).

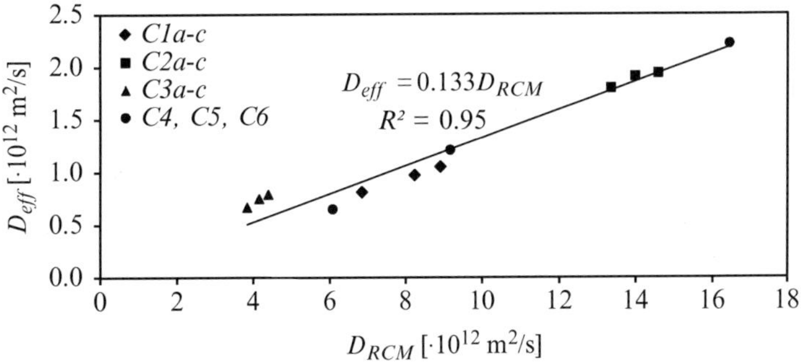


**Fig. 5** The effective chloride diffusion coefficient obtained from the extended model versus the chloride diffusion coefficient obtained from the basic RCM model

# 5 Conclusions

The following conclusions can be drawn from the present article:

- The chloride binding capacity, defined as $\partial c_b/\partial c$, is not constant in the concrete sample as implicitly assumed in the traditional RCM model. Due to the non-linear binding and non-equilibrium conditions between $c$ and $c_b$, the binding capacity changes locally with the chloride concentrations.
- The binding capacity during migration tests is very low at low free-chloride concentrations. Hence, the progression of the chloride penetration front through the concrete sample during migration tests is not retarded by chloride binding. Therefore, the $D_{RCM}$ as computed based on the position of the chloride penetration front is not affected as long as the chloride penetration depth in concrete is detected at very low free chloride concentration. Although this is the case for the $AgNO_3$ colourimetric chloride indicator for concretes based on OPC, it still needs to be verified for other types of binders.

## References


ASTM C 1556-4 (2004) Standard test method for determining the apparent chloride diffusion coefficient of cementitious mixtures by bulk diffusion. ASTM International, West Conshohocken

Atkinson A, Nickerson AK (1984) The diffusion of ions through water-saturated cement. J Mater Sci 19:3068–3078

Hooton RD, Thomas MDA, Stanish K (2001) Prediction of chloride penetration in concrete. US Department of Transportation, Federal Highway Administration FHWA-RD-00-142

Nilsson LO, Massat M, Tang L (1994) The effect of non-linear chloride binding on the prediction of chloride penetration into concrete structures. ACI Spec Publ 145:469–486

NT Build 443 (1995) Concrete, hardened: accelerated chloride penetration. Nordtest Method, Espoo

NT Build 492 (1999) Concrete, mortar and cement-based repair materials: chloride migration coefficient from non-steady-state migration experiments. Nordtest Method, Espoo

Spiesz P, Brouwers HJH (2012) Influence of the applied voltage on the rapid chloride migration (RCM) test. Cem Concr Res 42:1072–1082

Spiesz P, Ballari MM, Brouwers HJH (2012) RCM: a new model accounting for the non-linear chloride binding isotherm and the non-equilibrium conditions between the free- and bound-chloride concentrations. Constr Build Mater 27:293–304

Stanish K (2002) The migration of chloride ions in concrete. PhD thesis, University of Toronto

Tang L (1996) Chloride transport in concrete – measurement and prediction. PhD thesis, Chalmers University of Technology, Sweden

Yuan Q (2009) Fundamental studies on test methods for the transport of chloride ions in cementitious materials. PhD thesis, University of Ghent, Belgium

# Towards Structural Modelling of Alkali-Silica Reaction in Concrete

**Rita Esposito and Max A.N. Hendriks**

**Abstract** Structures affected by Alkali-Silica Reaction (ASR) include large massive concrete dams and bridges, which are considered essential for the infrastructure system, and not easily replaceable. The main feature of this reaction is the creation of a hydrophobic expansive gel which causes internal damage in the concrete. This damage is strongly related to the macroscopic stress state. Vice versa, the caused damage can have macroscopic consequences. Therefore a structural model should be able to capture the chemo-mechanical coupling induced by the swelling. For this purpose a multiscale material model is chosen to represent the ASR-affected concrete behaviour in structures. In this paper the motivations which brought the authors to this choice are explained.

**Keywords** Alkali-silica reaction (ASR) • Deterioration • Multiscale material modeling • Concrete structures

## 1 Introduction

The concrete infrastructure comprising bridges, dams and other civil works may be at risk due to deterioration caused by alkali-aggregate reaction. In this group of reactions the Alkali-Silica Reaction (ASR) is considered one of the most harmful processes, because it generates an expansive gel. This reaction, which begins at microstructural level, may eventually cause serious damage with consequent loss of structural capacity.

Due to the nature of the phenomenon, it should be studied from different points of view in order to understand the impact. For this reason, in 2010 at the Delft University of Technology the project Performance Assessment Tool for Alkali-Silica

R. Esposito (✉)
Department of Structural Engineering, Delft University of Technology, Delft, The Netherlands
e-mail: r.esposito@tudelft.nl

M.A.N. Hendriks
Department of Structural Engineering, Delft University of Technology, Delft, The Netherlands

Department of Structural Engineering, Norwegian University of Sciences and Technology (NTNU), Trondheim, Norway
e-mail: m.a.n.hendriks@tudelft.nl; max.hendriks@ntnu.no



C. Andrade et al. (eds.), *Durability of Reinforced Concrete from Composition to Protection*, DOI 10.1007/978-3-319-09921-7_16

Reaction (PAT-ASR) (Anaç and Esposito 2010) has been developed. The main goal is an understanding of the consequences of ASR by employing experiments and modelling approaches, both at different scales.

A large experimental campaign focussing on the characterization of a reference material with various tests method has been carried out. Meso-scale modelling of the experiments with the Delft Lattice Model (Anaç et al. 2012) is performed to investigate the reaction kinetics and the damage evolution. Meanwhile a material model for structural analysis (Esposito and Hendriks 2013; Esposito and Hendriks 2014) is under development in order to couple the chemical and mechanical loading action in ASR-affected concrete structures.

In this paper the idea and the motivations behind the formulation of the material model for the assessment of the structural effects induced by ASR are explained. An extensive literature review is included.

## 2 Experimental Observations

ASR is a chemical process between the alkali available in the cement and the silica originating from the aggregate. Its product is a hydrophobic gel which expands and builds up a pressure because it is confined by the concrete skeleton.

ASR has been studied by different researchers during several decades. Geologists and chemists investigated the mechanism of the gel formation; whereas civil engineers studied the mechanical impact both at material and structural level.

### *2.1 ASR Mechanism*

The chemical mechanism can be described by a two-stage process (Dent Glasser and Kataoka 1981). In the first stage the silica on the surface of the aggregates dissolves and reacts with the alkalis in the pore solution, thus creating the alkali-silica gel. In the second stage the gel comes in contact with the moisture and expands.

The development of the reaction depends on the type of aggregates involved and its crystal structure (Wang et al. 2004). In fact, the higher the structure's disorder of the reactive siliceous mineral phase, the faster it is its dissolution, so faster is the reaction. A well-crystallized material, such as sand quartz, is thermodynamically very stable. Its reaction with alkalis under normal conditions is extremely slow if not impossible. In contrast, a highly disordered reactive siliceous material as opal presents a fast reaction.

When the reaction is established, the gel can be observed in different parts of the microstructure. Microscopic investigations show the presence of the gel within the aggregates as well as in the cement paste and at their interfacial zone (Fig. 1). As reported by Saouma and Xi (2004) "Reaction initiates inside some selected (i.e. reactive) aggregates, gel forms, and in the presence of water swells. As it swells

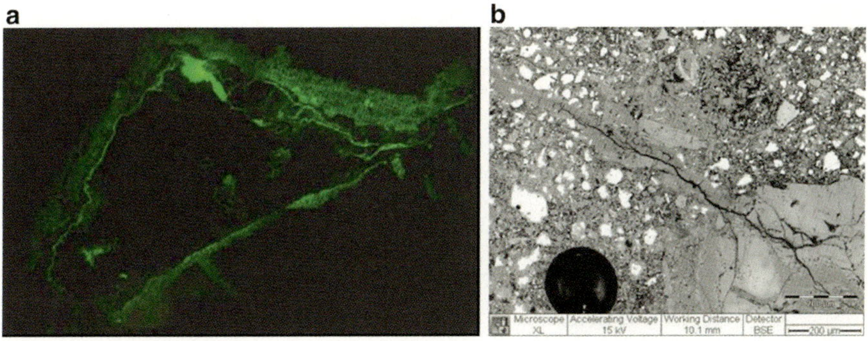


**Fig. 1** Microscopic observations: (**a**) Optical microphotographs of a damaged basaltoid aggregate and surrounding paste due to ASR (Çopuroğlu and Schlangen 2007); (**b**) penetration of ASR gel into the crack (Andic-Cakir et al. 2007)

in a confined environment, the aggregates eventually will crack to relieve the internal pressure and thus allow the gel to expand inside the newly formed void. One can speculate that the accumulated internal strain energy is much higher than the surface energy of the aggregate itself and that the excess energy drives the crack dynamically into the surrounding matrix and aggregates."

## 2.2 *Swelling of the Concrete Due to ASR Gel Expansion*

The expansion of ASR gel can lead to a macro swelling of the concrete with formation of cracks. However, before the gel expansion is visible the gel flows and fills the existing pore space, afterwards the internal pressure leads to microcracks formation and swelling of the concrete. The swelling appears to stop when either the alkali content is too low to react or the water does not reach the reactive site.

The swelling process can be influenced by the environmental conditions. Elevated temperatures can lead to a faster reaction (Fig. 2a). Besides, the moisture content is a relevant parameter. Laboratory tests have shown that if the relative humidity is lower than 50 % the concrete does not expand, even if the gel has been formed (Fig. 2b).

The swelling process is also influenced by the microstructure. Zhang et al. (1999) have reported that for aggregate between 0.15 and 10 mm, the smaller the aggregate size, the greater the concrete expansion. Multon et al. (2010) have shown that if the concrete contains only large reactive aggregates its expansion will be slower with respect to a concrete with different reactive aggregate sizes; this phenomenon can be explained with the difficulty of the alkali solution to enter the aggregates. Moreover when multiple aggregate sizes are considered the behaviour can be different on the basis of the ratio between the coarse and fine particles. This ratio determines how large the porous crown zone around the aggregates is: the larger the porous zone,

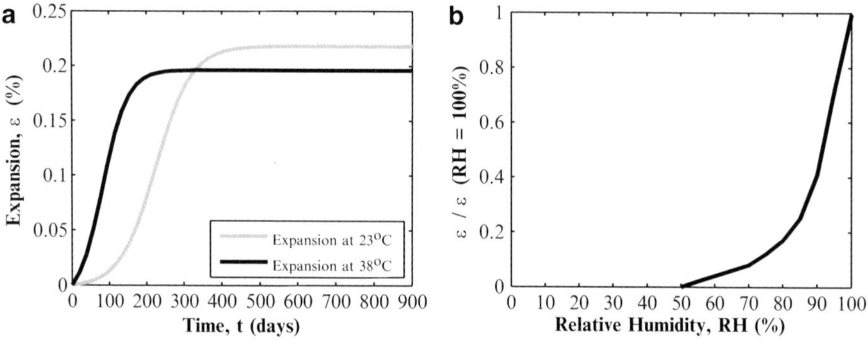


**Fig. 2** Influence of environmental conditions on concrete swelling: (**a**) Effect of temperature; (Larive 1998) (**b**) effect of relative humidity (Swamy 1992)

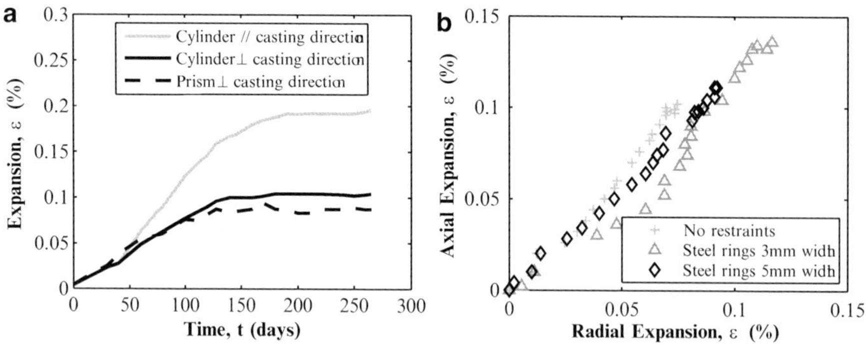


**Fig. 3** (**a**) Expansion curves for cylinders specimens and prisms specimens in free expansion conditions (Larive 1998), (**b**) expansions for specimens subjected to later constrains (Multon and Toutlemonde 2006)

the smaller the expansion because the gel has more space before pressurizing the concrete skeleton.

## 2.3 *Anisotropic Behaviour*

The swelling process of concrete affected by ASR appears to be characterized by an intrinsic anisotropic behaviour, as shown by Larive (1998). She observed that a sample in free expansion condition prefers to swell in the direction parallel to the casting direction; the expansion in this direction ranges from 1.3 to 2.8 times the expansion in the perpendicular directions (Fig. 3a).

Tensile tests on sound concrete specimens with the same aggregate size show, as well known, that the tensile strength is lower along the casting direction. This suggests that the distribution of pores with various shapes and orientations determines both the direction with the weakest tensile strength and the preferred expansion direction. Before microcracking occurs, the swelling is nearly isotropic. Afterwards the gel expansion will induce the propagation of the cracks in the weakest zone (perpendicular to the casting direction), which will mutually influence the further swelling. In conclusion, anisotropic cracking resulting from anisotropic strength properties influences the anisotropic expansion.

The same conclusion could be used to explain the swelling redistribution concept (Larive 1998; Multon and Toutlemonde 2006). In specimens subjected to compressive loading or lateral constraining the imposed expansion is lower in the restrained direction. Once again the gel expansion induces the crack propagation in the direction which requires less energy dissipation; when the constraints are applied in the lateral direction the gel tends to expand along the longitudinal one (Fig. 3b).

The anisotropic behaviour induced by the stress state is more relevant than the intrinsic one. This phenomenon is particularly important when the attention is focussed on the behaviour of ASR-affected concrete structures which can be seriously influenced by the coupling between chemical and mechanical loading and constraints.

## 2.4 Mechanical Degradation

The ASR reaction appears to degrade the mechanical properties of concrete differently than a mechanical loading (Esposito and Hendriks 2012). The earlier tests performed by Swamy and Al-Asali (1987) showed that: "the losses in engineering properties do not occur at the same rate or in proportion to the expansion undergone by the ASR-affected concrete". In literature, only few experimental campaigns are available on this topic (Larive 1998; Ahmed et al. 2003; Ben Haha 2006; Giaccio et al. 2008). They focus mainly on the compressive strength which appears not to be a good indicator for the estimation of the degradation, especially for low expansion values (Fig. 4a).

In Fig. 4 the degradation of the Young's modulus, the splitting tensile strength and the compressive strength is reported. The comparison is made for concrete

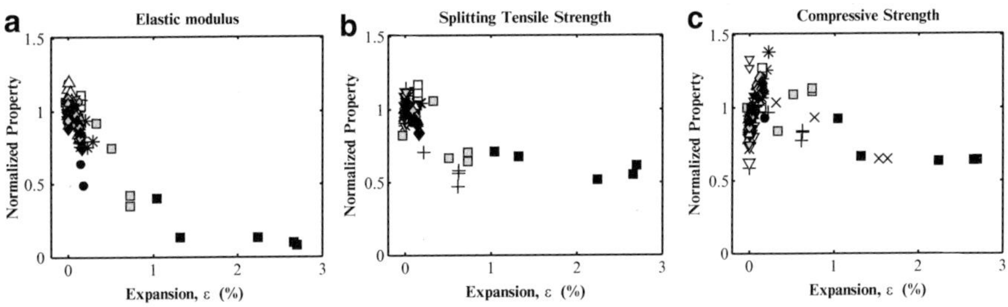


**Fig. 4** Degradation of mechanical properties in ASR-affected concrete samples in free expansion conditions: (**a**) Young's modulus; (**b**) splitting tensile strength; (**c**) compressive strength (Swamy and Al-Asali 1987, Larive 1998, Giaccio et al. 2008, Ahmed 2003, Ben Haha 2006)

samples stored in high humidity at a temperature between 28 and 40 ºC. The stiffness appears to be the most sensitive property; it can reach degradation levels up to 80 %. Moreover the stiffness and the strength degrade at a different rate.

## 3 Modeling ASR in Concrete

The alkali-silica reaction was first observed in large massive concrete structures as dams. In the early proposal a thermal equivalence approach was adopted to model the concrete expansion (Malla and Wieland 1999) and the stress-induced anisotropic behaviour was obtained by considering the local principal stresses (Charlwood 1994). In a second stage the influence of the environmental conditions was accounted by Leger et al. (1996) and Larive (1998) with phenomenological formulas.

Saouma and Perotti (2006) proposed an engineering approach that accounts for the swelling redistribution due to the stress state; they improved the method presented by Charlwood by introducing anisotropic expansion coefficients on the basis of experimental observations. Capra and Sellier (2003) adopted a probabilistic approach to model the evolution of cracks in the concrete matrix, due to the internal pressure generated by the swelling gel. The concrete is modelled like a damageable material having elastic and inelastic strains. ASR is modelled using global kinetics including temperature and humidity effects.

Recently, the attention has been shifted to describe the interaction between the gel and concrete skeleton at pore level, by investigating the kinetics of the phenomenon. Lattice models have been employed by Schlangen and Çopuroğlu (2010) and Anaç et al. (2012) to model the gel formation and its expansion. Concrete is modelled at meso-scale and the particles distribution is determined by image analysis of samples. Different expansion points are randomly selected into the micro-structures. Dunant and Scrivener (2010) adopted an extended finite element framework to perform micro-mechanical simulations of free expansion tests. They explain the damage induced by the ASR by introducing growing gel pockets in the aggregates.

Ulm et al. (2000) and Bangert et al. (2004) employed the porous media theory to describe the gel swelling and the development of the internal pressure as well as the chemomechanical coupling. Lemarchand et al. (2005) were able to describe the kinetics beyond the S-shaped expansion curve and to capture the swelling redistribution effect by simulating the cracks' closure (Lemarchand et al. 2003).

Furthermore some micro-mechanical models have been formulated to describe the "pessimum size" effect, to obtain more efficient expansion test procedures. This effect is strongly related to the aggregate size distribution and it defines the worst (pessimum) size which leads to the highest concrete expansion within a short time. Bažant et al. (2000) proposed a micro-mechanical fracture theory that explained the aggregate size effect in a 2-weeks accelerated test. Suwito et al. (2002), Poyet et al. (2007) and Multon et al. (2009) employed analytically-solved microscopic models to predict the size effect of the aggregates on the concrete expansion in a 1-year accelerated test.

# 4 PAT-ASR Project: Structural Modeling

In 2010 the PAT-ASR project (Performance Assessment Tool for Alkali-Silica Reaction) has been initiated at Delft University of Technology in order to study the main aspects of ASR in concrete. The aim is to provide a tool able to characterize the reaction and its damage both at material level and in the sense of structural response.

The micro-mechanical aspects of ASR have a complex interaction with the macro-mechanical state of the concrete, therefore, following the work of Lemarchand et al. (2003) and Charpin and Ehrlacher (2012), a multiscale approach is adopted (Fig. 5). The properties of concrete are evaluated on the basis of micro-mechanical aspects and the behaviour of a Representative Elementary Volume (REV) is analysed (Fig. 6a).

The sound concrete is modeled as an heterogeneous material (Fig. 6) composed by aggregates and microcracks embedded in the cement paste. Each material is behaving elastically. The aggregates are modeled as spheres. The microcracks are modelled by three orthogonal families of penny-shaped inclusions.

The alkali-silica reaction is simulated by changing the microstructure. The chemical process starts at aggregate level by consuming the silica available and forming the gel, which is modelled as spherical inclusions into the aggregates (Fig. 6b). It is assumed that the gel has a volume bigger than the volume of the eroded aggregate, therefore a pressure is generated. The gel flows into the cement paste by filling the microcracks around the aggregates. Eventually the pressure is so high to generate damage in the system.

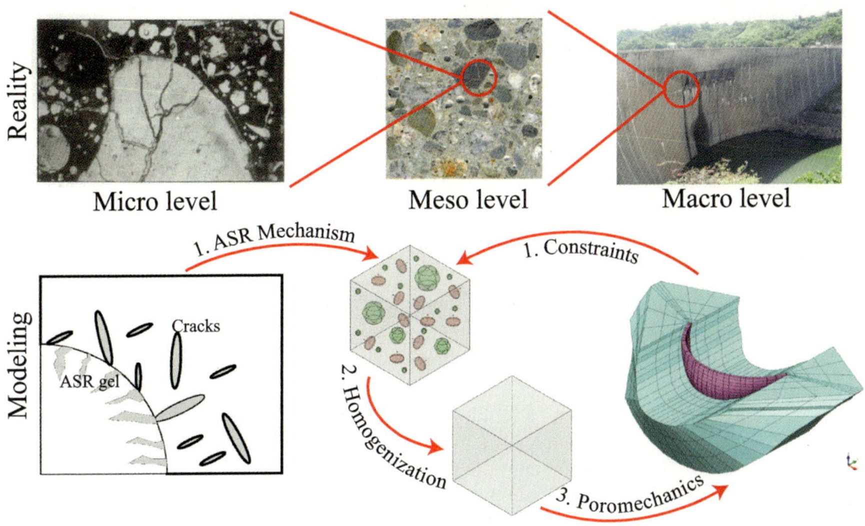


**Fig. 5** Modelling procedure for structural analyses

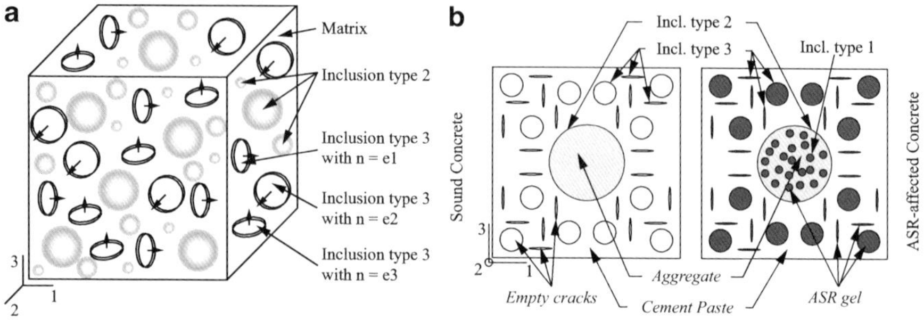


**Fig. 6** Micro-mechanical model: (**a**) 3D representation; (**b**) sound and ASR-affected concrete

The damage evolution is formulated in the framework of linear fracture mechanics by employing an energy-based damage criterion. Both the external mechanical loads and the internal pressure contribute to the description of the damage, which leads to an increase of the crack radii. The effective properties of the medium are analytically evaluated by the Mori-Tanaka homogenization scheme.

The model accounts for the micromechanical changes provoked by the ASR swelling. It is able to describe the stress-induced anisotropy effect of ASR, as well as the degradation of the mechanical properties resulting from the combined effect of chemical and mechanical loading (Esposito and Hendriks 2013; Esposito and Hendriks 2014). The coupled effect of chemical and mechanical loading is seen as a key point for explaining differences between results from laboratory tests and the observed behaviour of ASR-affected structures.

By employing this approach a more fundamental model is adopted, which is able to capture the micro and macro aspects of the problem. The ASR-affected concrete is seen as an evolving material, whose state should be followed over time taking into account chemical and mechanical loading conditions.

Even based on a simplified microstructure, which allows analytical homogenization, the model turns out to predict well the behaviour of both sound and ASR-affected concrete and to achieve the goal of modelling the chemomechanical coupling. An analytical homogenization is preferred in order to facilitate its implementation in a finite element program. An extensive description of the model and its validation will be presented in an upcoming publication.

This approach allows employing the outputs of the model in a more straightforward approach to be used in practice. Moreover, the model can be helpful to interpret and supplement laboratory tests, which are usually requested to assess the material behaviour of an existing concrete structure suffering from ASR.

Eventually, since this approach goes to the basis of the interaction between chemical and mechanical loads, it could also be employed to describe related degradation mechanisms in concrete, e.g. concrete internally pressurized by steel corrosion products or concrete subjected to water ingress.

**Acknowledgements** This work is part of the project "Performance Assessment Tool for Alkali-Silica Reaction" (PAT-ASR, http://pat-asr.blogspot.nl/), which is developed in the context of the IS2C program (http://is2c.nl/).

The authors wish to express their thanks to the Dutch National Foundation (STW), the Dutch Ministry of Infrastructures and the Environment (Rijkswaterstraat), SGS and TNO DIANA BV for their financial support.

## References

Ahmed T, Burley E, Rigden S, Abu-Tair AI (2003) The effect of alkali reactivity on the mechanical properties of concrete. Constr Build Mater 17(2):123–144

Anaç C, Esposito R (2010) Development of a performance assessment tool for alkali silica reaction (PAT-ASR). http://pat-asr.blogspot.nl/

Anaç C, Schlangen E, Çopuroğlu O (2012) Lattice model implementation on alkali silica reaction gel expansion in a reacted concrete medium. In: The 3rd international conference on concrete repair, rehabilitation and retrofitting (ICCRRR-3), Cape Town

Andic-Cakir O, Copuroglu O, Schlangen E, Garcia-Diaz E (2007) Combined experimental and modelling study on the expansions of concrete microbars due to ASR. In: CONSEC07, Tours

Bangert F, Kuhl D, Meschke G (2004) Chemo-hygro-mechanical modelling and numerical simulation of concrete deterioration caused by alkali-silica reaction. Int J Numer Anal Methods Geomech 28(7–8):689–714

Bažant ZP, Zi G, Meyer C (2000) Fracture mechanics of ASR in concretes with waste glass particles of different sizes. ASCE J Eng Mech 126(3):226–232

Ben Haha M (2006) Mechanical effects of ASR in concrete studied by SEM-image analysis. Laboratoire des matériaux de construction. PhD thesis, École Polytechnique Fédérale de Lusanne (EPFL), Lausanne

Capra B, Sellier A (2003) Orthotropic modelling of alkali-aggregate reaction in concrete structures: numerical simulations. Mech Mater 35(8):817–830

Charlwood RG (1994) A review of alkali aggregate in hydro-electric plants and dams. Hydropower Dams 5:31–62

Charpin L, Ehrlacher A (2012) A computational linear elastic fracture mechanics-based model for alkali-silica reaction. Cem Concr Res 42(4):613–625

Çopuroğlu O, Schlangen E (2007) Modelling of effect of ASR on concrete microstructure. Key Eng Mater 348–349:809–812

Dent Glasser LS, Kataoka N (1981) The chemistry of 'alkali-aggregate' reaction. Cem Concr Res 11(1):1–9

Dunant CF, Scrivener KL (2010) Micro-mechanical modelling of alkali–silica-reaction-induced degradation using the AMIE framework. Cem Concr Res 40(4):517–525

Esposito R, Hendriks MAN (2012) A Review of ASR modeling approaches for finite element analyses of dam and bridges. In: 14th international conference on alkali aggregate reaction, Austin

Esposito R, Hendriks MAN (2013) Multiscale material model for ASR-affected concrete structures. In: The XII international conference on computational plasticity. Fundamentals and Applications (COMPLASXII), Barcelona

Esposito R, Hendriks MAN (2014) Modeling of alkalis-silica reaction in concrete: a multiscale approach for structural analysis (submitted). In: Proceedings of the computational modeling of concrete and concrete structures (EURO-C), St. Anton am Alberg

Giaccio G, Zerbino R, Ponce JM, Batic OR (2008) Mechanical behavior of concretes damaged by alkali-silica reaction. Cem Concr Res 38(7):993–1004

Larive, C. 1998. Apports combinés de l'expérimentation et de la modélisation la comprehénsion de l'alcali-réaction et de ses effets mécaniques. PhD thesis, Laboratoire Central des Ponts et Chaussées (LCPC), Paris

Leger P, Cote P, Tinawi R (1996) Finite element analysis of concrete swelling due to alkali-aggregate reactions in dams. Comput Struct 60(4):601–611

Lemarchand E, Dormieux L, Kondo D (2003) A micromechanical analysis of the observed kinetics of ASR-induced swelling in concrete. In: Bicanic N, DeBorst R, Mang H, Meschke G (eds) Computational modelling of concrete structures. A A Balkema Publishers, Rotterdam, pp 483–490

Lemarchand E, Dormieux L, Ulm FJ (2005) Micromechanics investigation of expansive reactions in chemoelastic concrete. Philos Trans R Soc Math Phys Eng Sci 363(1836):2581–2602

Malla S, Wieland M (1999) Analysis of an arch gravity dam with a horizontal crack. Comput Struct 72:267–278

Multon S, Toutlemonde F (2006) Effect of applied stresses on alkali-silica reaction-induced expansions. Cem Concr Res 36(5):912–920

Multon S, Sellier A, Cyr M (2009) Chemo–mechanical modeling for prediction of alkali silica reaction (ASR) expansion. Cem Concr Res 39(6):490–500

Multon S, Cyr M, Sellier A, Diederich P, Petit L (2010) Effects of aggregate size and alkali content on ASR expansion. Cem Concr Res 40(4):508–516

Poyet S, Sellier A, Capra B, Foray G, Torrenti JM, Cognon H, Bourdarot E (2007) Chemical modelling of alkali silica reaction: influence of the reactive aggregate size distribution. Mater Struct 40(2):229–239

Saouma V, Perotti L (2006) Constitutive model for alkali-aggregate reactions. Aci Mater J 103(3):194–202

Saouma V, Xi Y (2004) Report CU/SA-XI-2004/001 – literature review of alkali aggregate reactions in concrete dams (Original work published in 10 Nov 2004)

Schlangen E, Copuroğlu O (2010) Modeling of expansion and cracking due to ASR with a 3D lattice model. In: Fracture mechanics of concrete and concrete structures (FraMCos7). Korea Concrete Institute, Seoul

Suwito A, Jin W, Xi Y, Meyer C (2002) A mathematical model for the pessimum size effect of ASR in concrete. Concr Sci Eng 4(13):23–34

Swamy RN (ed) (1992) The alkali-silica reaction in concrete. Van Nostrand Reinhold, New York

Swamy RN, Al-Asali MM (1987) Engineering properties of concrete affected by alkali-silica reaction. ACI Mater J 85:367–369

Ulm FJ, Coussy O, Li KF, Larive C (2000) Thermo-chemo-mechanics of ASR expansion in concrete structures. J Eng Mech ASCE 126(3):233–242

Wang H, Farzam H, Bryan S (2004) ASR – a review of mechanisms and proper prevention measurements. In: The 12th international conference on alkali-aggregate reaction, Beijing

Zhang C, Wang A, Tang M, Wu B, Zhang N (1999) Influence of aggregate size and aggregate size grading on ASR expansion. Cem Concr Res 29(9):1393–1396

# Free Expansion Tests for ASR at the Level of a Single Aggregate-Matrix Interface: Experimental Results and Numerical Modelling

**Joaquín Liaudat, Carlos M. López, and Ignacio Carol**

**Abstract** As part of an on-going research project, a model being developed to simulate ASR in concrete is described. The formulation consists of the three main diffusion/reaction field equations for the concentrations of Alkali, Calcium and Silicate ions in the pore solution, complemented by a number of evolution equations for the local ingredients involved (e.g. local concentration of reactive silica, local availability of portlandite, etc.).

A new experimental set-up to study the expansion mechanisms at the level of a single matrix-aggregate interface, using sandwich-like cylindrical specimens of cement paste/mortar and glass are exposed to 1 M NaOH solution at 60 °C, is also presented along with typical free expansion curves and subsequent SEM images and EDX analysis of the reaction.

**Keywords** Alkali-silica reaction • Concrete • Interface expansion • Reactive transport • Numerical modeling

## 1 Introduction

Alkali-Silica Reaction (ASR) involves a set of simultaneous chemical reactions between chemical species in the concrete pore solution and certain types of metastable silica present in the aggregates. These reactions are relatively well known (Dron and Brivot 1992, Wang and Gillott 1991, Hou et al. 2004), and can be used to explain several aspects of the expansions caused by ASR, such as the dependence on the alkali content of the concrete, the importance of relative humidity or the kind of potentially reactive minerals.

However, there are some issues of ASR, such as the anisotropy of expansions due to ASR in concrete under triaxial stress states (Multon and Toutlemonde 2006) that cannot be explained only on the basis of the chemical reactions involved, and compel

J. Liaudat (✉) • C.M. López • I. Carol
ETSECCPB (School of Civil Engineering), UPC (Technical University of Catalonia), Barcelona, Spain
e-mail: joaquin.liaudat@upc.edu; carlos.maria.lopez@upc.edu; ignacio.carol@upc.edu

 
C. Andrade et al. (eds.), *Durability of Reinforced Concrete from Composition to Protection*, DOI 10.1007/978-3-319-09921-7_17

to consider the Chemo-Transport-Mechanic coupling underlying ASR expansion. To the best of the authors' knowledge no mathematical model of this kind has been reported. A model of these characteristics, properly validated with experimental data, would make it possible to assess, at least qualitatively, the real influence of mitigation measures on concrete affected by ASR from a holistic point of view.

In this sense, the ongoing research project aims at formulating a coupled chemo-transport-mechanic reaction mechanism of ASR expansions in concrete and implementing it in a FE model, as it has been done before by the research group for external sulfate attack (Idiart et al. 2011).

In this paper, only general aspects of the formulation under development and part of the ongoing experimental campaign are presented.

## 2 Numerical Model for ASR

### *2.1 Reaction Mechanism and Chemical Reactions*

The reaction mechanism proposed is summarized in the scheme shown in Fig. 1. The interfacial zone between hardened cement paste (HCP) and aggregate in concrete is represented. Both components are assumed to be porous and fully saturated with water at all times during the reaction. Immediately after casting the HCP has a certain content of portlandite and its pore water has a certain content of calcium and alkali ions. Sodium and potassium ions are treated indistinctively in this paper as alkali ions. The aggregates are supposed to have a certain content of reactive silica and no calcium or alkali ions in their initial pore solution.

The concentration gradient of alkali and calcium ions causes a diffusion process from the cement paste towards the aggregate, which is followed by an increase of the pH of the pore solution in the aggregate. This increase causes the dissolution of the reactive silica through silicate ions, which in turn, due again to a concentration gradient, diffuse towards the cement paste. At the same time, these three diffusion processes drive the dissolution or precipitation of portlandite.

Wherever, in the aggregate or in the cement paste, the three reactants (calcium, alkali and silicate ions) are available a reaction occurs forming a

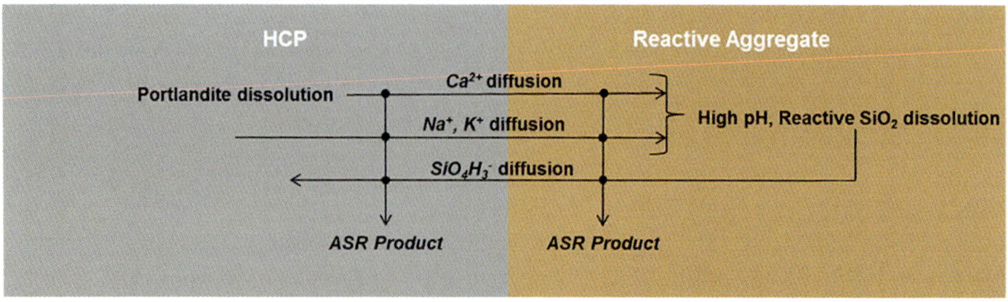


**Fig. 1** Scheme of ASR reaction mechanism

Calcium-Alkali-Silicate-Hydrate (ASR gel) of variable composition, depending on the concentration of the reactants. The composition, in turn, determines the molar volume of this product.

The variability of the composition is reached by considering two separate reactions, each one with a fixed stoichiometry and its own kinetic law, one reaction forming an Alkali-Silicate-Hydrate (ASH) and the other forming a Calcium-Silicate-Hydrate (CSH). Adding both products we have the expansive ASR gel.

Finally, at a given location, the volume balance of the solid constituents, namely reactive silica, portlandite and ASR gel, determine the transport properties and the volume of pore solution. If the volume of precipitated ASR gel is greater than the available space in capillary pores plus the additional space liberated by silica and portlandite dissolution, a localized volumetric expansion may occur.

### 2.2 Formulation

The reaction mechanism is formulated in terms of the concentrations of the different chemical species in the pore solution. The three diffusion processes are considered by means of Fick's second law, with sink/source terms given by the chemical kinetics of the silica dissolution and ASR gel formation. The concentrations of hydroxyls, portlandite, reactive silica and ASR gel are considered by means of local variables, given by the same set of chemical kinetic equations. Finally, two further restrictions are imposed by the equilibrium equations of water self-ionization and portlandite dissociation.

The volumes of the solid constituents at given location are obtained by multiplying the associated concentration variables by their molar volumes. In that way, two opposite processes may be simulated simultaneously. On one hand the porosity is increased by silica and portlandite dissolution and, on the other hand, it is reduced by ASR gel precipitation. If the volume of ASR gel is greater than the available empty space in the capillary pores, a macroscopic expansion is produced. Likewise, the changes in the volumetric fractions of the solid constituents (in particular in the capillary porosity) modify the diffusion coefficient of the material.

More details of the formulation and its numerical implementation in 1D may be found in (Liaudat et al. 2013).

## 3 ASR Free Expansion Tests at the Level of a Single Aggregate-Matrix Interface

### 3.1 Test Setup

The aim of this test is to measure the expansions at the level of a single interface aggregate-matrix due to the precipitation of ASR products. For this purpose, two kinds of cylindrical specimens of 33 mm diameter and 66 mm height are elaborated. The first type, named "Active specimens", involves sandwich–like specimens, with cement paste or mortar on top and bottom of a disc of reactive aggregate in the middle. The second one, called "Control specimens", involves purely mortar or cement paste specimens without reactive aggregate. Control specimens are used to assess the deformations of the matrix of cement paste or mortar, solely caused by phenomena such us drying shrinkage or thermal expansions. The alkali content of the cement pore solution is increased by adding NaOH to the mixing water.

So far, soda-lime glass and borosilicate glass have been used as reactive aggregates because of their advantageous of homogeneity as industrial materials. Once the methodology had been demonstrated, natural reactive aggregates will be used.

After casting, the specimens are kept in the molds for 24 h in a humid chamber. Once unmolded, they are cured in airtight containers with a solution of sodium hydroxide of 1 mol/l at 23 °C during 27 days. Finally, the containers are introduced in an oven at 60 °C until the end of the test. The sodium hydroxide solution keeps the specimens saturated of water and prevents the leakage of alkali ions from the cement paste. During both stages, specimen length changes are measured regularly with a dial indicator until the end of the test.

It must be mentioned that a similar test has been previously proposed by Schlangen and Çopuroglu (2007). However, they have not presented, to the best of the author's knowledge, any results since then.

## *3.2 ASR Products at the Glass-HCP Interface After Test*

Once in the oven at 60 °C, the active specimens start to expand and to expel a whitish reaction product at the interfaces, until the moment when the amount of reaction product at the interface is high enough to separate the glass from the matrix, which can happen in a matter of days, weeks, or months depending on the dosage (mainly on NaOH added and sand content) of the matrix and the glass type.

Immediately after the separation the reaction products in the interface form a continuum lamina with a waxen texture (Fig. 2, left). A few minutes later, exposed to the air they get dry and take a scaly appearance (Fig. 2, right).

Before the separation, the reaction products at the interface are studied by means of SEM images and EDS analysis of polished sections of the specimens. For example, in Fig. 3 a set of images of the interfacial zone of an active specimen of cement paste and soda-lime glass can be seen. In the left image, the upper part correspond to the glass, the bottom part to the hardened cement paste (HCP) at and the middle, a region of about 100 μm width, to the reaction products. The reaction products can only be seen at the borders of this zone, just beside the HCP and the glass. The rest is plain black due to the epoxy used in the preparation of the sample. It is assumed that the reaction products should have been there but due to the cutting and polishing processes they have been lost. In the upper-right image, the

**Fig. 2** Reaction products at the glass-cement paste interface of active specimens once separated; immediately after taking them out of the 1 M NaOH solution at 60 °C (*left*) and a few minutes later exposed to the air (*right*)

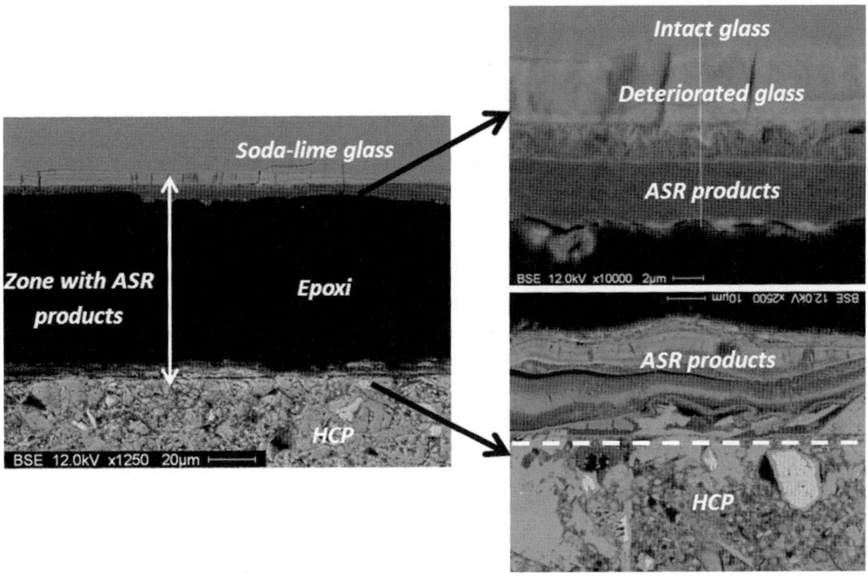


**Fig. 3** SEM images from the interfacial zone of an active specimen of cement paste and soda-lime glass after exposure to a 1 M NaOH solution at 60 °C

degradation process of the glass can be appreciated in detail, from the intact glass to the ASR product. In the bottom-right image, the original position of the interface can be clearly distinguished as a straight line separating the hardened cement paste from the ASR products.

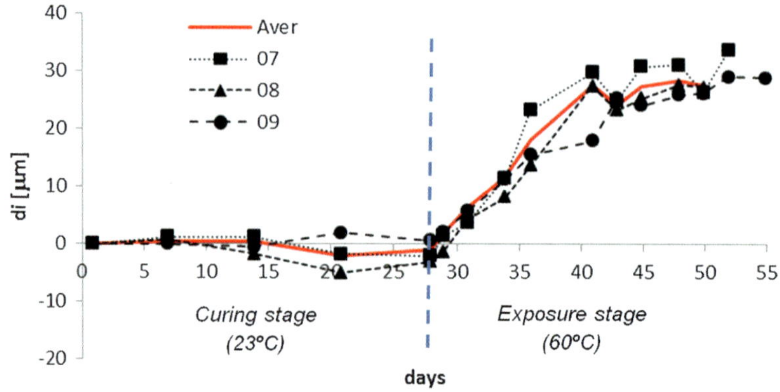


**Fig. 4** Interfacial expansion curves for three identical specimens of cement paste and soda-lime glass

### 3.3 Expansion Curves

From the length change measurements of active and control specimens, the expansion curves corresponding to the ASR reaction products precipitated in a single interface matrix-aggregate is obtained using Eq. 1, where $L$ and $\Delta L$ are the length and the length change of the active specimen respectively, $e$ is the width of the glass disc, $\varepsilon^c$ is the average deformation of the control specimens, $\alpha_{glass}$ is the thermal expansion coefficient of the glass, $\Delta T$ is temperature variation from curing (23 °C) to exposure (60 °C) and $d_I$ is the expansion at a single interface.

$$d_I = \frac{1}{2}\left[\Delta L - \overline{\varepsilon}^c (L - e) - e \propto_{glass} \Delta T\right] \quad (1)$$

A typical set of interfacial expansion curves corresponding to specimens of cement paste and soda-lime glass is presented in Fig. 4. When the curing stage is finished and the temperature is increased to 60 °C the sudden change of the slope, from practically zero at 23 °C to 2 μm per day, can be appreciated. After 12 days at 60 °C the curves reached a plateau before one of the interfaces spontaneously separated. The maximum expansions measured in this case range between 20 and 30 μm per interface.

## 4 Concluding Remarks

Considering a chemo-transport coupled mechanism for ASR would allow reproducing some aspects of the phenomenon which have not been reproduced in previous models. The most interesting aspects are:

- The variability of the composition of the reaction product, both in space and time.
- The dependence of the swelling capacity on the composition of the reaction products.

- The effect of portlandite availability on the silica dissolution and, therefore, on the extent of ASR.
- The effect of portlandite and silica dissolution and the precipitation of reaction products on the overall diffusivity coefficients.

A new experimental test is being developed to validate this reaction mechanism by means of a 1D numerical model. Interesting information about the reaction products and the parameters affecting ASR can be obtained from these tests at a very low cost.

Next steps in this project include:

- Validating the presented diffusion–reaction model with experimental data from the literature and from interfacial expansion tests.
- Implementing the proposed diffusion–reaction mechanism in a coupled scheme with a meso-mechanical model (DRAC) for simulating concrete, and
- Setting up interfacial expansion tests under vertical load to assess the influence of the stress state on the expansions.

**Acknowledgements** This research is supported by grants BIA-2012-36898 funded by MEC (Madrid), which includes FEDER funds, and 2009SGR-180 from AGAUR-Generalitat de Catalunya (Barcelona). The first author also wishes to acknowledge MICINN (Madrid), for his FPI doctoral fellowship.

The authors thank A. Leemann for the SEM images presented in this paper.

## References

Dron R, Brivot F (1992) Thermodynamic and kinetic approach to the alkali-silica reaction. Part 1: Concepts. Cem Concr Res 22:941–948

Hou X, Struble LJ, Kirkpatrick RJ (2004) Formation of ASR gel and the roles of C-S-H and portlandite. Cem Concr Res 34:1683–1696

Idiart AE, López CM, Carol I (2011) Chemo-mechanical analysis of concrete cracking and degradation due to external sulfate attack: a meso-scale model. Cem Concr Compos 33:411–423

Liaudat J, López CM, Carol I (2013) Diffusion–reaction model for alkali-silica reaction in concrete. In: Proceedings of the XII international conference on computational plasticity, Barcelona

Multon S, Toutlemonde F (2006) Effect of applied stresses on alkali–silica reaction-induced expansions. Cem Concr Res 36:912–920

Schlangen E, Çopuroglu O (2007) Concrete damage due to alkali-silica reaction: a new method to determine the properties of expansive gel. In: Carpinteri A (ed) Fracture mechanics of concrete and concrete structures – high-performance concrete, brick- masonry and environmental aspects. Taylor & Francis Group, London, pp 1835–1841

Wang H, Gillott J (1991) Mechanism of alkali-silica reaction and the significance of calcium hydroxide. Cem Concr Res 21:647–654